# Modeling and Verification Using UML Statecharts

# Modeling and Verification Using UML Statecharts

## A Working Guide to Reactive System Design, Runtime Monitoring and Execution-Based Model Checking

Doron Drusinsky

AMSTERDAM • BOSTON • HEIDELBERG • LONDON
NEW YORK • OXFORD • PARIS • SAN DIEGO
SAN FRANCISCO • SINGAPORE • SYDNEY • TOKYO

Newnes is an imprint of Elsevier

Newnes is an imprint of Elsevier
30 Corporate Drive, Suite 400, Burlington, MA 01803, USA
Linacre House, Jordan Hill, Oxford OX2 8DP, UK

**Library of Congress Cataloging-in-Publication Data**

Drusinsky, Doron.
  Modeling and verification using UML statecharts : a working guide to reactive system design, runtime monitoring, and execution-based model checking / Doron Drusinsky.
      p. cm.
  ISBN 0-7506-7949-2 (pbk. : alk. paper)   1. UML (Computer science)  2. Formal methods (Computer science) 3. Computer software--Development.  I. Title.
  QA76.76.D47D78 2006
  005.1'17--dc22

                              2006005265


**British Library Cataloguing-in-Publication Data**

A catalogue record for this book is available from the British Library.
ISBN-13: 978-0-7506-7949-7
ISBN 0-7506-7949-2


For information on all Newnes publications,
visit our website at www.books.elsevier.com.

06 07 08 09 10   10 9 8 7 6 5 4 3 2 1
Printed in the United States of America.

# Dedication

*This book could not have been written without the lifelong support of my parents Harry and Luba.*

*To my wonderful children, Dana, Gabi, Shiron, and Maya with whom I learned the real meaning of reactive systems, and to my beloved wife Dganit.*

# Contents

# Preface

In my twenty years of practice in the software and computer science field, I was fortunate enough to have two careers: one in the industry and one in the academia. Wearing these two hats I witnessed successful transitions of research to commercial applications, such as in the cases of cryptology and digital signal processing (DSP).

Formal methods, an assortment of mathematical methods for the specification, development, and verification of software, did not enjoy such a success. After being researched for a quarter of a century or more by some of the most brilliant minds in the world, formal methods have been adopted in a very limited manner by the industry.

From an academic perspective, the most common explanation for this lackluster acceptance is that the *problem is hard.* In other words, the problem that academic research usually tries to address—mathematically proving that a program conforms to a formal specification—is a hard problem to solve using computer-aided tools due to computer science complexity-theory related issues. I refer to this problem as the *verification problem.*

From an industry perspective, however, the core issues seem rather different. Engineers and programmers want techniques that reduce their pain or win them a gold mine, and hopefully both. It is therefore hard to sell to engineers and programmers the idea that some unknown academic—albeit mathematical—formal specification language is actually better in capturing requirements than simply coding in them directly in Java. The idea of having yet another language one needs to master, resulting in three separate views of the component that need to be maintained and synchronized (formal specification, source code, and UML) is hard to sell without having a clear benefit as an end goal—a benefit that's hard to justify given the verification problem discussed above. In short, the prime issue seen from the industry is about specification. I call this problem the *specification problem*.

To be truthful, some formal methods have been recently accepted by the software industry; specifically, these are methods that relate to the specification and verification of *transformational components* (the distinction between transformational and reactive systems is described in Chapter 2). Techniques such as design-by-contract, manifested by the Java Modeling Language (JML), are now used by many Java developers. In fact, you may think of this book as suggesting corresponding techniques and tools suitable for *reactive components*.

This book addresses the specification problem first and foremost. The book describes UML statecharts, the primary UML language when it comes to reactive components. It then describes how to use the same diagrammatic language for specifying *requirements* for reactive components (which we call *temporal* requirements) instead of using a special academic language. Having both the component design and its formal requirement specification done in the same language highlights the primary question engineers have always asked about formal methods approach, namely: *why bother*? Why not just have one kind of statechart—the design state-chart? This is an excellent question and I devote an entire chapter to it.

The book also addresses the verification problem using run-time monitoring, a lightweight method that is admittedly not perfect but it works and scales for real systems. I then show how to extend run-time monitoring with automatic test generation for the purpose of constructing an execution-based model checker.

# Acknowledgments

I am indebted, far and foremost, to Dr. D. Caffal of the MDA for his vision and courage without which many of the ideas presented in this book would not have come to light.

My colleagues at NPS, specifically Bret Michael, Man-tak shing and Tom Cook have provided me with excellent feedback and support throughout the last couple of years. Members of the MDNT, including Scott Pringle, Nick Sklavounos, Dion Hinchcliffe, Chris Kauffman, Erik Stein, Steve Apsel, Dirk Penberthy, Thad Goodwyn, and Craig Trader have all provided me with excellent feedback over the course of the last two years. Mike Robison provided excellent writing and editing support. Last but not least Dganit Drusinsky assisted me to bring this material to the light of day.

# What's on the CD-ROM?

CD-ROM contains:

CD-ROM contains diagrams and code for the example in Chapter 4, and also includes runtime code for monitoring and automatic white box test generation.

# Chapter 1

# Formal Requirements and Finite Automata Overview

## 1.1. Terms

Many students consider the theory of finite automata and formal languages theoretical and irrelevant to their future livelihood. Indeed, the theory is more often than not taught as a prelude to complexity theory.

In this book we will put a fresh spin on the theory, using it as a prelude to UML-based modeling, specification, and verification of reactive systems. To be relevant to reactive systems in general, and to UML in particular, we will use domain-specific terms, listed in Table 1.1, that are not usually associated with formal languages.

TABLE 1.1 UML vs. Formal-Language terms.

| Terms Used in This Book | Classical, Formal-Language Counterparts |
|---|---|
| Domain of Discourse | Alphabet(s) |
| Event or Condition | Alphabet letter |
| Scenario | String |
| Specification or Requirement | Formal Language |

Throughout this chapter we will consider using automata and formal languages in the context of the specification or design of a software component, which we will call the *component under design*.

## 1.2. Finite Automata: The Basics

### 1.2.1. The Domain of Discourse (Alphabet)

Formally speaking, an *alphabet*, typically represented with the Greek letter $\Sigma$, is a finite set of symbols called *letters*. In practice, these symbols are the names of events or conditions in the domain of discourse for the component under design. For the sake of simplicity, and to be able to tie our discussion closely to the theory of formal languages, in this chapter we will mostly interpret alphabet letters as events. We will leave the distinction between conditions and events to Chapter 2, where we will see how statecharts accommodate both.

A question often asked is, *Which events do we include in the alphabet?* The answer is simple: *every* event we might need for modeling or specification. In other words, the domain of discourse, as its name suggests, contains all the events that need to be taken into account during those design phases. It is therefore important to nail down the domain of discourse before proceeding to the modeling or specification phases. All subsequent modeling and specification will be based on the domain of discourse.

Consider, for example, a traffic-light controller that receives the following inputs from its environment: *oneMinuteElapsed*, *newCar*, and *newAmbulance*. The alphabet for the controller is then $\Sigma_{tnl} =$ {*oneMinuteElapsed*, *newCar*, *newAmbulance*}.

For alphabets that consist of conditions, there are two approaches to the relationships among member conditions.

The first, taken by formal language theory, says that conditions (letters) in a single alphabet are by definition always mutually exclusive. More precisely, exactly one condition from the alphabet must be true at any given time. If two conditions could be true simultaneously, they must be associated with distinct alphabets. Hence, a system with three unrelated conditions $C_1$, $C_2$, and $C_3$ has a domain of discourse that consists of the three alphabets $\Sigma_1 = \{C_1, !C_1\}$, $\Sigma_2 = \{C_2, !C_2\}$, and $\Sigma_3 = \{C_3, !C_3\}$.

The second approach, the one we will use in the context of statecharts in Chapter 2, says that all conditions are unrelated, so that any condition can be true at any given time. This amounts to the creation of a distinct alphabet, $\Sigma_C$, for every condition, C, where $\Sigma_C = \{C, !C\}$.

Events are always considered as pair-wise mutually exclusive for reasons we will discuss in Chapter 2.

## 1.2.2. An Input Scenario (String)

Since they are mutually exclusive, input events arrive one at a time as inputs to the component under design. Such a sequence of inputs is called a *scenario*, also known as a *string*. In other words, a scenario is a sequence of alphabet events. Consider, for example, the following alphabet with two events, $\Sigma = \{open, close\}$. The sequences *open.close.open* and *open.open.open.close.close* are two scenarios. Note how the *sequencing operator*, a period or point, "." (also known as the *concatenation* operator) is used to represent the order of events in a scenario, such as event *close* following event *open* in the scenario *open.close*.

For historical reasons theoreticians consider an input string to reside on a device called the *input tape*. In this book, however, we will not consider events to reside anywhere. In fact, we will assume that the events are lost forever after being received and processed by the component under design.

A scenario event induces a *cycle* in the component under design. Hence, a scenario of length 100 (a sequence of 100 events) induces 100 cycles in the component under design. The length of a scenario, *seq,* is denoted $|seq|$.

The empty scenario, denoted $\varepsilon$, is one that contains no events. It is useful mostly for mathematical purposes, such as to construct slick proofs by mathematical induction or to create recursive definitions. Obviously, since $\varepsilon$ contains no events, $\varepsilon.x = x.\varepsilon = x$ for every scenario.

The symbol $\Sigma^*$ denotes all possible finite scenarios that can be constructed from the events of the alphabet $\Sigma$. The * operator is known as the *Kleene* star operator. Note that the empty scenario is considered a member of $\Sigma^*$ whereas $\Sigma^* - \{\varepsilon\}$ is denoted $\Sigma+$. $\Sigma^*$ represents all possible scenarios that can be constructed with the events of the domain of discourse.

## 1.2.3.  A Requirement (A Formal Language)

A *requirement*, also known as a *formal language*, is a *set of scenarios* constructed from events of a given domain of discourse, $\Sigma$. In other words, any subset of $\Sigma^*$ constitutes a requirement. As software component developers and designers, we are not interested in just any subset of $\Sigma^*$ but rather in specific ones. Intuitively, therefore, we can see that a requirement is a specification of *legal scenarios for a component under design*. We will discuss this interpretation in Chapter 4.

There are many ways a human being can describe a requirement, such as using Natural Language (NL) or set notation. Consider, for example, a requirement, R1.2.3a, using the alphabet $\Sigma$ = {*open, close*} that consists of those scenarios that *never contain two consecutive open events or two consecutive close events*. Obviously, this description is the natural language version of the requirement. Another example is the requirement R1.2.3b, using the same alphabet, which consists of the two scenarios {*open.close, close.open*}. This description clearly uses set notation.

As we will see in Chapters 3 and 4, natural language is ambiguous and currently not well understood by computers. Set notation is problematic to use for requirements, like R1.2.3a, that consist of an infinite number of scenarios. Therefore, we need a formalism that is precise and unambiguous, that is easy to use and readable by developers and customers, and that at the same time can be understood or executed by a computing device. That is precisely the challenge this book addresses.

Note that interesting requirements are almost always infinite; that is, they contain an infinite number of scenarios. Constituent scenarios, however, are sequences of finite length. Finite requirements are usually uninteresting because they are so limited that they do not contribute much to the overall specification of a system. Consequently, in this book, we will not consider methods used to specify finite requirements, such as UML message sequence charts (MSCs).

Consider two related requirements:

1. R1.2.3a.

2. R1.2.3c, defined using the alphabet of events $\Sigma$ = {*open, close*}, consisting only of those scenarios that never contain two consecutive *open* events.

Clearly, R1.2.3c ⊂ R1.2.3a. It is therefore tempting to think of R1.2.3a as a requirement that captures R1.2.3c. That would be incorrect for the following reason. When we specify a requirement we do so for both *legal* and *illegal* scenarios. Stated differently, by specifying those scenarios that are legal for a component under design we are indirectly also specifying that all other scenarios are illegal. If, however, we consider R1.2.3a as a requirement that captures requirement R1.2.3c, whatever is illegal for R1.2.3a would also be illegal for R1.2.3c, whereas the scenario *close.close* (two consecutive *close* events) is actually *legal* for R1.2.3c.

Requirements can be sequenced (concatenated). Given requirements $R_1$ and $R_2$, for example, the requirement $R_1.R_2$ consists of all scenarios we can construct as r1.r2, where $r_1$ is in $R_1$ and $r_2$ is in $R_2$.

The Kleene star operator applies to requirements. Given a requirement R, the requirement R* consists of all scenarios we can construct by concatenating scenarios from R to one another any number of times (including zero). In other words, R* = {ε} ∪ R ∪ R.R ∪ R.R.R ∪ R.R.R.R … For example, if R = {*open.close, close.close*}, all of the following scenarios are members of the requirement R*: ε, *open. close, close.close, open.close.close.close, close.close.open.close, open.close.open.close.open.close, open.close.close.close.open.close. close.close*. On the other hand, *open.open* is not a member of R* because there is no way to construct it by repeatedly concatenating elements of R.

Clearly, unless R is the empty set, R* consists of infinitely many scenarios. For example, if R = {*open*}, R* = {ε, *open, open.open, open.open.open,* …}.

Thus, the Kleene star operator provides a formal way to specify infinite sets of scenarios using set notation. For example, R1.2.3a can

now be written as {open.close}* ∪ {*open.close*}*.{*open*} ∪{*close.*
*open*}* ∪ {*close.open*}*.{*close*}

## 1.2.4. A Specification

A (formal) specification is a representation of a requirement using notation a computer can understand and read in a finite amount of time using finite resources. For example, a table that lists all member scenarios for a requirement is not an acceptable specification because the requirement might be infinite. Similarly, with the current state-of-the-art technology, a natural language description of a requirement is not an acceptable specification because, for the most part, a computer cannot understand it.

The primary goal of this book is to introduce a visual, familiar, and intuitive specification language based on UML statecharts.

## 1.3. Regular Expressions

Regular expressions (REs) are a substitute for set notation used to specify requirements as sets of scenarios. With REs, the standard set-notation braces are stripped off, and the set union operator (∪) is written as a +. Hence, requirement R1.2.3a is captured by the RE:

(*open.close*)* + (*open.close*)*.*open* + (*close.open*)* + (*close.open*)*.
*close*

## 1.4. Deterministic Finite Automata and Finite State Diagrams

Finite automata (FA)[1] have a textual definition and a graphical counterpart called finite state diagrams. We will use the latter throughout the book as a representative of FA.

An FA is a graph with a finite number of vertices, called *states*. Edges of a finite state diagram are called *transitions* and are annotated with symbols from the domain of discourse, $\Sigma$. In addition, one of the states must be the (unique) initial state, and any number of states, from 0 to the total number of states n, are chosen as final states (visually depicted with a circle inside a circle), as shown in Figure 1.1a.

Note that the domain of discourse, $\Sigma$, must be defined first, and only then can we define the FA. Nevertheless, we will omit the detailed specification of $\Sigma$ from most of our diagrams because it can be deduced from the transition labels, as in Figure 1.1a where clearly $\Sigma = \{p, q\}$.

Given a state $S$, of an FA, a *next-state of S for the input p*, denoted NS($S, q$), is a state pointed to by a transition that is issued from $S$ and is labeled with $q$. In Figure 1.1a, for example, $B$ is a next-state of $A$ for input $p$.[2]

---

[1]   FA also stands for the singular, "finite automaton."

[2]   In traditional automata theory, this relationship is often written $\delta(S, q)$.

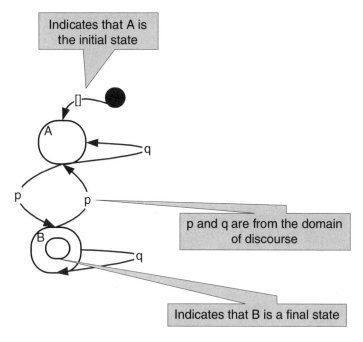

a. A fully-specified DFA.

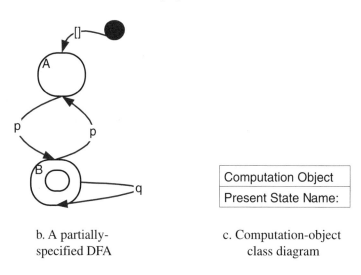

b. A partially-
specified DFA

c. Computation-object
class diagram

FIGURES 1.1a–c.

### 1.4.1.  Fully and Partially Specified Deterministic Finite Automata

An FA is a *deterministic* FA (DFA) if NS($S$, $q$) is unique. A DFA is *fully specified* if NS($S$, $q$) is always defined and *partially specified* otherwise. For example, the FA in Figure 1.1a is deterministic and fully specified and the FA in Figure 1.1b is deterministic and partially specified because NS($A$, $q$) is undefined. A fully specified DFA is denoted an FDFA.

### 1.4.2.  A Computation (a Run)

A DFA $D$ performs a *computation*, also known as a *run*, as a response to an input scenario from $\Sigma^*$ (the same $\Sigma$ used to define the DFA). The computation that $D$ performs is defined as the sequence of states, starting from the initial state, that $D$ visits while traversing transitions as it reads the symbols of the scenario one by one. In other words, the DFA computation is the sequence of states along a continuous path that the DFA traverses, starting in the initial state, while reading one input symbol at a time and traversing a transition annotated with that symbol.

For example, the FDFA in Figure 1.1a performs the computation *comp* = A.B.A.A.B.B in response to the input scenario $seq_1$ = p.p.q.p.q. Specifically, it performs the state changes illustrated in the pseudo-animation in Figure 1.2 (traversed transitions and resulting computation states are highlighted with thick lines).

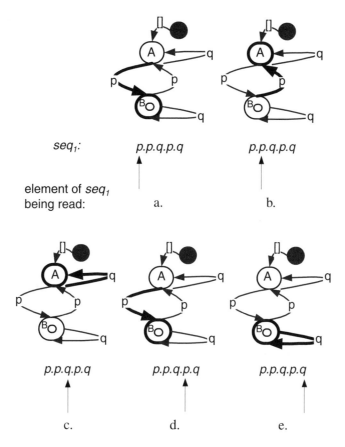

FIGURES 1.2a–e   Pseudo-animation of a DFA response for the DFA
of Figure 1.1a and input scenario *seq₁*.

By definition, a DFA responds to a given input scenario with at
most one computation. A fully specified DFA always responds to an
input scenario with some computation, because it always has a well-
defined next state to visit no matter where it is and no matter what
the current input symbol is. A partially specified DFA, however,
might not be able to respond to certain input scenarios. For example,
the DFA in Figure 1.1b cannot respond to the scenario $seq_1$ because
when in state *A* it has no specified next-state for the input *q*.

This definition of a response for a partially specified DFA differs from the definition used by statecharts (we will take up statecharts in Chapter 2). For the latter, an unspecified next-state $NS(S, p)$ is considered to be $S$ itself; in other words, the statechart simply *stutters* in the same state if no transition can be traversed. Hence, the machine in Figure 1.1b responds to the scenario $seq_1$ with the computation *comp* because it assumes that $NS(A, q) = A$.

### 1.4.3. An Object-Oriented Interpretation of a Computation

A convenient way to think about a computation is to use an *object-oriented interpretation* as follows. A computation consists of a list of objects, called *computation-objects*, that visit states of the DFA. A computation-object, whose class diagram is shown in Figure 1.1c, holds the identity of the state being visited in a variable called the *present-state*. The computation response of the DFA is then a sequence of computation-objects. The first computation-object in the computation holds the identity of the initial state, $S_0$, in its present-state; the computation-object is said to be *visiting* $S_0$ and is denoted $<S_0>$. When the DFA reads the first symbol, $s_1$, of the input scenario, $<S_0>$ constructs a child computation-object that visits the state $S_1$, reachable from the present state via a transition annotated with $s_1$; this computation object is denoted $<S_1>$. Hence the computation at this point is the sequence $<S_0>.<S_1>$. When the DFA reads the second symbol $s_2$ of the input scenario, $<S_1>$ constructs its child computation-object that visits state $S_2$, reachable from the present state $S_1$ via a transition annotated with $s_2$. Hence the computation now is the sequence $<S_0>.<S_1>.<S_2>$. In other words, the last computation-object of the computation sequence creates a child object when the DFA reads a symbol from the input scenario; the child is then appended to the end of the computation. Each time a new child object is created is called a *cycle*.

A DFA computation is a *linear* sequence (a chain) of computation-objects because when the last computation-object of a computation is <*S*> and the input symbol being read is *s*, there can be at most one transition issuing from *S* that is annotated with *s* or else the FA is not deterministic.

DFA computation-objects are trivial because the only item they encapsulate is the present-state variable. Later, we will extend this definition to include additional data while maintaining the same intuitive approach.

Hence the object-oriented computation that the DFA in Figure 1.1a creates in response to the input scenario $seq_1$ is the list, *comp* = <A>.<B>.<A>.<A>.<B>.<B>.

A partially specified DFA might not always be able create a new computation object. Consider, for example, the response of the partially specified DFA in Figure 1.1b to input scenario *p.q.q*. After it reads the first *p* symbol, the computation is <A>.<B>. In the next cycle, however, the DFA has no specified transition issuing from B that is labeled with *q*. The last computation-object (<B>) is then said to *die,* and therefore the computation dies too.

## 1.4.4. Specifying Requirements Using DFA

Recall that a requirement is a set of scenarios (those we consider legal). DFA are used to describe requirements as follows.

First, we define how a DFA, *D*, accepts or rejects a scenario, *s*. Say *comp* is *D*'s computation in response to *s*. If *comp* ends with a final state, the DFA is said to *accept s*; if it doesn't, it is said to *reject s*. In other words, the scenario *s* is *legal*, or *good*, if and only if *comp* ends with a final state; otherwise it is *illegal*, or *bad*. Note that if comp does not exist, as is sometimes the case for partially specified DFA, $s$ is considered to have been rejected by the DFA. For example, the partially specified DFA in Figure 1.1b rejects the scenario $seq_1 = p.p.q.p.q$ because it performs no computation in response to $seq_1$.

The requirement *specified*, *described*, or *captured* by a DFA is the set of scenarios that the DFA accepts. For example, the DFA in Figure 1.1a captures the requirement whose natural language description is:

**R1.4.4**: *a scenario is legal if and only if it contains an odd number of p symbols.*

### 1.4.5. Practical Issues

You are probably wondering about two practical issues:

1.  Where are the symbols of the input scenario generated?

2.  How does the DFA know when to visit a state or, said differently, what if the DFA tries to read a symbol from the input before that symbol is available?

A superficial answer would be, *This is a theoretical chapter and your questions are about the implementation.* A more accurate answer, however, is that they are actually good observations that go to the heart of the matter, as follows. There are two common types of input symbols, events and conditions, which we will discuss in greater detail in Chapter 2. Think of events as generated by the outside world—i.e., after the DFA receives an event it is lost. In contrast, think of conditions as Boolean variables within the same system as the DFA that the DFA can access any time. The arrival of an event, then, is the trigger that tells the DFA to traverse a transition. In other words, the DFA does not *actively* read an input symbol; rather, it is *passively* told (by the event) to read the symbol, in fact *the event is the symbol.* This type of computation is called *event-driven,* and most of our discussion in Chapter 2 will be related to such computations. In the case of conditions, because the DFA can access the input symbols any time, it can visit states as fast as we wish. This type of computation is called *procedural,* to be discussed in Section 2.7.

## 1.4.6. Regular Requirements

A requirement is *regular* if it can be described by a DFA. This definition gives us a handle on the descriptive power of DFA for comparison purposes. For example, the specification language of linear-time temporal logic (LTL), which we will examine in Chapter 3, is not capable of specifying every regular requirement or, stated differently, there exist one or more requirements that LTL cannot specify but a DFA can.

## 1.5. Nondeterministic Finite Automata

A *nondeterministic FA* (NFA) is a partially or fully specified FA in which, for some state $S$, and for some input symbol $p$ it is legal for the next-state of $S$, [NS($S$, $p$)], *is not necessarily unique.* In other words, a state $S$ could exist in which two issuing transitions have the same label but different target states. This situation is called a *next-state conflict.*

Figure 1.3a shows an NFA. Clearly the next-state of A0 for input symbol $p$ is not unique: it can be either A0 or A1, corresponding to the two issuing transitions from A0 labeled with $p$.

Just as a DFA responds to an input scenario with a computation, an NFA responds to an input scenario with a *computation tree,* as follows. Computation objects are defined as they are for DFA—namely, they hold the identity of a state being visited. Unlike the case with the DFA object-oriented interpretation, when an NFA reads an input symbol $s$ while visiting a state $S$, it might see two or more transitions issuing from $S$ that are annotated with $s$. For example, in Figure 1.3a, there are two possible successor computation-objects to <A0> for the input symbol $p$, <A0> and <A1>. To resolve this, computation-object <A0> creates two successors, <A0> and <A1>. Let's continue this example with the input symbol $p$. <A1>'s successor for the input $p$ does not exist. <A1>'s successor for the input $p$ is as before. Let's continue this example with the input symbol $q$. Computation-object <A1> now creates <A4> and <A2> as its successors, and <A0> creates <A4> as its successor. Consequently, the response of the NFA to the input scenario $seq_2 = p.p.q.q.q$ is the computation tree shown in Figure 1.3b.

The *leaves* of the computation tree response to $seq_2$ are defined as the nodes on the bottom of the tree. Put another way, a node must be of height $|seq|$ to be considered a leaf. This definition is such because an NFA is potentially partially specified, as in Figure 1.3a, and indeed the computation object <A1> dies when the NFA reads the input symbol $p$.

A leaf is said to be *accepting* when it ends with a computation-object that represents a final state, like leaf <B>.

A different view of the computation tree is as a collection of linear computations, called *computation paths*, which are similar to DFA computations. For example, the computation tree in the figure contains, among others, the computation path <A0>.<A0>.<A1>.<A4>.<A4>.<A4> and <A0>.<A0>.<A1>.<A2>.<A3>.<B>. Each path behaves like a DFA computation, in effect "believing" that it is the "chosen" computation. We will use this approach when we consider the behavior of nondeterministic Java statecharts in Chapter 3.

Note that every DFA is by definition a special case of an NFA in which no next-state conflicts exist.

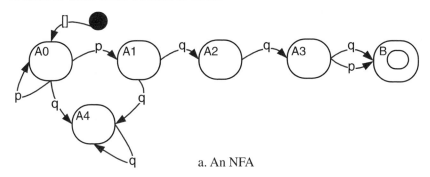

a. An NFA

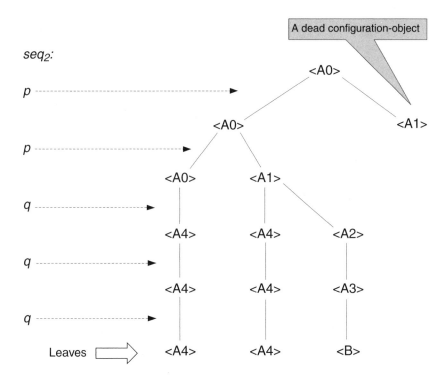

b. The computation tree response of the NFA
of (a) to the input scenario $seq_2 = p.p.q.q.q$

FIGURES 1.3a–b An NFA and its response to input scenario $seq_2$.

## 1.5.1. Specifying Requirements Using NFA

An NFA, *N*, accepts a scenario, *seq*, if and only if its computation-tree response to *seq* contains an accepting leaf. For example, the NFA in Figure 1.3a accepts $seq_2$. Stated differently, the NFA accepts *seq* if a computation-path exists that ends in an accepting leaf. Consequently, a scenario, *seq*, is rejected by *N* if and only if *all* leaves in its computation-tree response to *seq* do not represent a final state.

As for DFA, the requirement *specified, described,* or *captured* by an NFA is the set of scenarios that the NFA accepts.

If we think of a state diagram for an NFA as a nondeterministic statechart (as we will do in Chapter 3), computation-objects never die, but stutter (as discussed in Section 1.4.2). For example, childless <A1> computation-object will actually stutter in <A1> as long as *p* is the input symbol.

## 1.5.2. NFA "Guessing"

It is common to describe the way an NFA accepts a scenario as *guessing*. This kind of guessing is not the intuitive stochastic kind we might expect—using a coin toss for example. Rather, what is meant is that instead of a computation object in the computation tree creating more than one successor, it picks a single successor. The chosen child is one that is on an accepting path—that is, a path that ends with an accepting leaf. Obviously, there is no practical way to implement this guessing other than to create all child objects and to try all computation-paths. Nevertheless, the term guessing is convenient as shorthand for the more elaborate description that involves multiple computation-paths.

## 1.6. Other Forms of FA

### 1.6.1. Universal ∀-FA

Universal-FA, abbreviated as ∀-FA, look and feel just like NFA. The only difference is the definition of acceptance for an input scenario.

Recall how the definition of scenario acceptance for an NFA is intrinsically *existential*; that is, a scenario is accepted by an NFA if a leaf computation-object in its computation-tree response *exists* that represents a final state. A ∀-FA uses a dual definition where a scenario, *seq*, is accepted if and only if *all* the leaves of the computation-tree response to seq correspond to the final states.

Just as for DFA and NFA, the requirement *specified*, *described*, or *captured* by a ∀-FA, *F,* is the set of scenarios accepted by *F*.

### 1.6.2. Alternating Finite Automata

Alternating FA (AFA) combine NFA and ∀-FA using two types of states. Hence, they look like DFA, NFA and ∀-FA but with one difference: AFA states are marked as either *and* (&&) states or *or* (‖) states, as in Figure 1.4a.

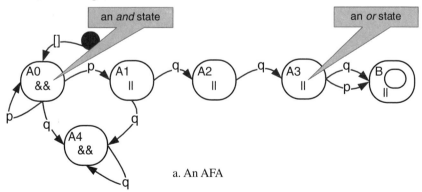

a. An AFA

FIGURE 1.4a   An AFA, its computation-tree response, and computation-tree traces.

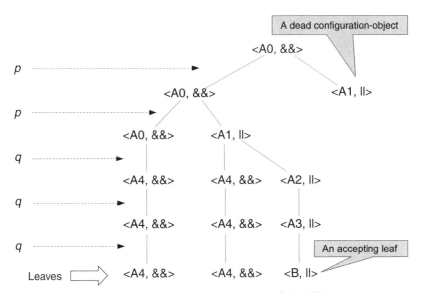

b. The computation-tree response of the AFA
to the input scenario $seq_2 = p.p.q.q.q.$

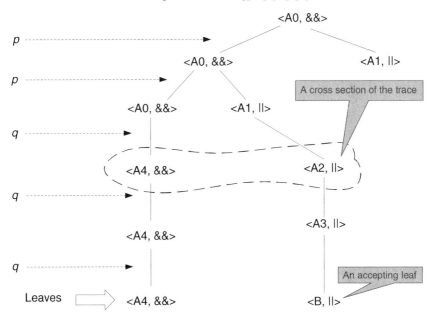

c. A trace of the computation tree of (b)

FIGURES 1.4b and c.

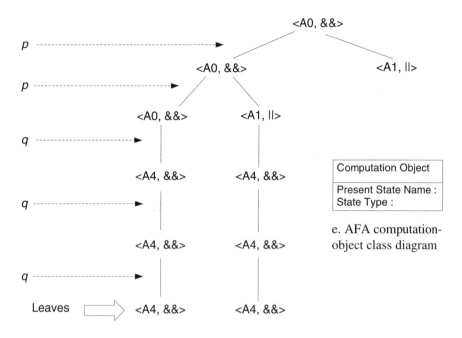

d. Another trace of the computation tree of (b)

FIGURES 1.4d and e.

AFA computation-objects carry a state type in addition to a state identity, as shown in the class diagram in Figure 1.4e. An AFA state can either be an *and* (&&) state or an *or* (||) state. Other than that, an AFA computation tree is created just as an NFA computation tree is. Figure 1.4b shows the computation tree response for the AFA in Figure 1.4a to the input scenario $seq_2 = p.p.q.q.q$.

A computation tree *trace* is a sub-tree of the computation tree in which:

1. For every *and*-type computation-object in the trace, *all* its computation-tree child objects are also in the trace.

2. For every *or*-type computation-object in the trace, *one* of its computation-tree child objects is also in the trace.

Figures 1.4c and 1.4d show two traces of the computation tree in Figure 1.4b, which is the response of the AFA in Figure 1.4a to $seq_2$

A trace is similar to a $\forall$-FA computation tree. For example, a trace is said to be *accepting* if *all* its leaves are accepting. The two traces shown are not accepting because both have some nonaccepting leaves.

An AFA *A* accepts a scenario *s* if the computation tree response of *A* to *s* contains an accepting trace. None of the traces in Figure 1.4 are accepting, for example, neither is any other trace, and therefore the AFA (Figure 1.4a) rejects $seq_2$.

Here, too, the requirement specified, described, or captured by an AFA is the set of scenarios accepted by *A*.

Note the relationships among the four types of FA. Every DFA is by definition a special case of an NFA in which no next-state conflicts exist. Every NFA is by definition a special case of an AFA in which all states are *or*-type states. And every $\forall$-FA is by definition a special case of an AFA in which all states are *and*-type states.

AFA are an inconvenient language for human beings to use to specify requirements. Nevertheless, they are useful for algorithmic and analysis purposes.

### 1.6.3. NFA, $\forall$-FA and AFA with ε-Transitions

Consider the following NL requirement, which describes a certain kind of binary number.

**R3.3**: *A number is legal if and only if it contains one or two digits and possibly a '–' prefix.*

The ε-NFA in Figure 1.5a—so-called because it uses a special label ε on some of its transitions—realizes this requirement. The ε label is a new, artificial symbol outside the domain of discourse; that is, it is not the name of an event or condition. Whenever a computation-object visits a state that issues ε-transitions, like Init or B, it immediately creates a child computation object for the next state. Hence, the ε-NFA shown boots up with a single computation-object, <Init>, but immediately creates <A>. Similarly, when it creates the computation-object <B> it immediately creates the child computation object <OK>.

Recall that given an input scenario, $S$, the computation tree responses of an NFA, ∀-FA, and AFA are the same. The only difference is in the contents of the computation-objects (AFA objects include a type) and the definition of scenario-acceptance. Therefore, we can apply ε transitions to ∀-FA and AFA as well, along with the same object-oriented behavior as for ε-NFA.

The benefit of ε-transitions is that sometimes they simplify specifications. For example, the ε-NFA in Figure 1.8a is arguably more readable than the equivalent NFA in Figure 1.8b. In addition, ε-transitions enable compositional specifications. Suppose, for example, that we have already created two NFA specifications for requirements $Ra$ and $Rb$, and we want to create a specification for the requirement $Rc = Ra \cup Rb$, that is, $Rc$ contains all scenarios that are in either $Ra$ or $Rb$. Figure 1.8c shows the simple ε-NFA specification for $Rc$. Likewise, Figure 1.8d shows the simple ε-∀-NFA specification for $Rc = Ra \cap Rb$.

Note that ε-transitions automatically create nondeterministic behavior. In Figure 1.8a, for example, there is no obvious next-state conflict. Nevertheless, once the NFA boots up it immediately has more than one computation-object. Similarly, whenever the computation-object <B> is created, so is <OK>.

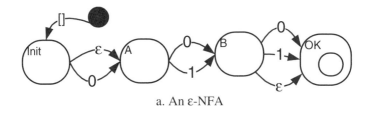

a. An ε-NFA

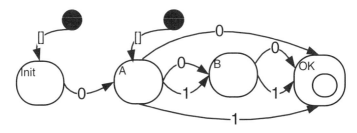

b. The equivalent of NFA

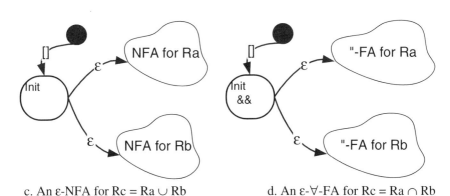

c. An ε-NFA for Rc = Ra ∪ Rb                    d. An ε-∀-FA for Rc = Ra ∩ Rb

FIGURES 1.5a–d   ε-NFA and ε-∀-FA.

## Flowcharting vs. ε-Transitions

In Chapter 2 we will describe flowcharts within statecharts. Flow-chart transitions resemble ε-transitions in that they are instantly tra-versed, without waiting for an external event. Nevertheless, they differ from ε-transitions in one important aspect: they do not induce

automatic nondeterminism. In other words, a deterministic statechart can include a flowchart whereas a DFA with ε-transitions is
an oxymoron, because an ε-transition automatically induces more
than one computation object.

## 1.6.4. Multitape FA

A multitape FA is an FA in which the domain of discourse alphabet
is the composition, or Cartesian product, of several "smaller," independent alphabets. For example, consider a traffic-light controller
(TLC) DFA that receives input symbols from two entities, sensors
and a network. The sensor-related symbols are represented by the
alphabet $\Sigma_1$ = {*itsACar, itsATruck, itsAPedestrian*}, and the network alphabet events by the alphabet $\Sigma_2$ = {*itsNetworkSyncMode,
itsIndependentMode*}. Clearly, any FDFA specification for the TLC
has six transitions issuing from every state, one for every pair of
possible input symbols, such as $A \rightarrow <_{itsACar,\ itsIndependentMode}>B$, which
means that the transition $A \rightarrow B$ is traversed if and only if the current
input symbols are *itsACar and itsIndependentMode* from $\Sigma_1$ and
$\Sigma_2$. In other words, every such FDFA uses an implied alphabet, $\Sigma$
= $|\Sigma_1| \times |\Sigma_2|$.

Consequently, FDFA are seldom used for the specification
of multitape FA. DFA specifications for the multitape TLC use
the following convention. A transition, $A \rightarrow_{itsACarB}$, for example, is
shorthand for the transitions $A \rightarrow <_{itsACar,itsIndependentMode}>B$, $A \rightarrow <_{itsACar,}$
$_{itsNetworkSyncMode}>B$.

Note that there can be only a single alphabet of events, for reasons we will discuss in Chapter 2.

## 1.7.   FA Conversions and Lower Bounds

### 1.7.1.  From NFA and ∀-FA to DFA

In this section we will see that every NFA, $N$, has an equivalent DFA, $D$; namely, some $D$ can specify every requirement specified by $N$. We can create an equivalent $D$ so that it mimics $N$, using a construction known as the *subset construction.*

Consider again Figure 1.3b, which shows the computation tree response of the NFA in Figure 1.3a to the input scenario $seq_2$. Because the nodes of the tree are computation-objects, we can encapsulate all objects created in one cycle inside a new kind of computation-object called a *compound-object,* whose class diagram is shown in Figure 1.6a. For example, the first input $p$ results in two computation objects, <A0> and <A1>, as shown in Figure 1.3b, so we encapsulate that pair in a single compound-object, <A0, A1>. Similarly, we represent the child objects that result from the third input $q$ as compound-object <A4, A2>, as in Figure 1.3b.

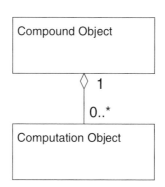

FIGURE 1.6a   Compound object class diagram.

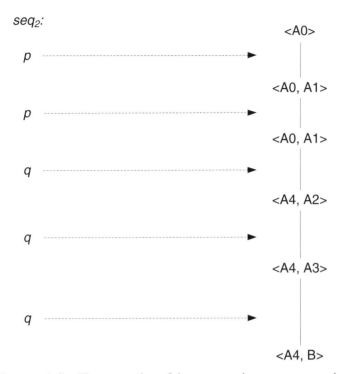

*seq₂:*

<FIGURE 1.6b  The conversion of the computation tree response in
Figure 1.3b into a linear computation.>

FIGURE 1.6b   The conversion of the computation tree response in
Figure 1.3b into a linear computation.

A compound computation-object, like <A4, A2>, is just a wrapper that encapsulates several individual computation-objects. The wrappers make an NFA computation tree (as in Figure 1.3b) a linear computation, which is a legal DFA computation (as in Figure 1.6b). Hence the DFA $D$ we will construct will indeed be a DFA that responds to $seq_2$ with this linear computation.

We construct $D$'s states so that they correspond to compound configuration-objects. In other words, we construct a graph vertex in $D$'s state diagram for every possible compound configuration-object.

We construct $D$ using the following *discovery* technique, shown in Figure 1.7: First, we define $D$'s initial state, <A0>, as the compound object that contains $N$'s initial computation object (Figure 1.7a). We say that we've *discovered* <A0>.

- Next, we repeatedly discover the transitions issuing from the previously discovered states. For example, after we have discovered the state for compound object <A4,A3> (Figure 1.7d), the next step is to observe the behavior of the original NFA, N, when it visits computation objects <A4> and <A3>. N has transitions A4$\rightarrow_q$A4 issuing from A4, and A3$\rightarrow_q$B and A3$\rightarrow_p$B, issuing from A3. Hence, N creates the computation-object <B> when it sees the input symbol $p$, and the computation-object <A4,B> when it sees the input symbol $q$. Therefore the DFA must include the corresponding compound-object transitions: <A4,A3>$\rightarrow$p<B> and <A4,A3>$\rightarrow_q$<A4,B> (Figure 1.7e).

- The process stops when no new transitions can be added to $D$.

Note the state labeled $\emptyset$ in Figure 1.7. It represents the situation in which $N$ has no live computation objects.

$D$'s final states consist of all compound-objects in which *at least one member* computation-object represents a final state in $N$, such as <A4,B> and <B>. This definition mimics $N$'s *existential* acceptance criteria.

In the worst case, $D$'s state space could be large. Because $N$ has $n$ possible computation objects, $n$ being the number of states in $N$, $D$ potentially has $2^n$ possible states, one for every subset of $N$'s states.

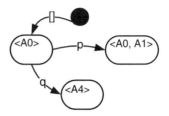

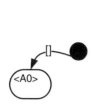

a. First step of
discovery: create
*D*'s initial state.

b. Second discovery step:
create transitions issuing
from the initial state,
discover resulting states.

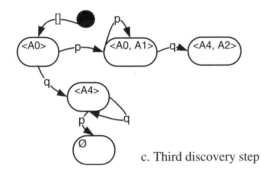

c. Third discovery step

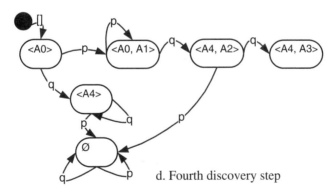

d. Fourth discovery step

FIGURES 1.7a–d   Discovering a DFA, *D*, that
mimics the NFA in Figure 1.3a.

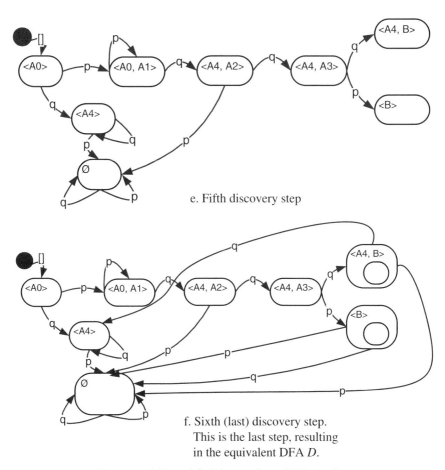

e. Fifth discovery step

f. Sixth (last) discovery step.
This is the last step, resulting
in the equivalent DFA $D$.

FIGURES 1.7e and f   Discovering a DFA $D$, that
mimics the NFA in Figure 1.3a.

The subset construction technique applies to the conversion of
$\forall$-FA into DFA as well, the only difference being the definition of
$D$'s final states, which now consist of all compound-objects in which
*all* member objects represent a final state in $N$.

The subset construction shows that NFA (and $\forall$-FA) are capable of specifying only *regular requirements*. Nevertheless, for certain regular requirements they are exponentially more succinct
than DFA.

## 1.7.2. From ε-NFA to NFA

The conversion of ε-NFA into NFA is straightforward. We replace every path in an ε-NFA that ends in a chain of ε's with a single transition, as shown in Figures 1.8a and 1.8b, where A→$_1$ B→$_ε$ OK is replaced with A→$_1$ OK. Also, if there are ε transitions issuing from the initial state, we create multiple initial states accordingly.

The existence of this conversion means that ε-NFA, too, are only capable of specifying *regular requirements* because every ε-NFA can be converted into an equivalent NFA, which in turn can be converted into a DFA.

## 1.7.3. From Regular Expressions to ε-NFA

The conversion of regular expressions into ε-NFA is likewise straightforward. In Section 3.9.2 we will discuss a more general conversion of extended regular expressions into nondeterministic statecharts.

## 1.7.4. From AFA to NFA or ∀-FA

The conversion of AFA into NFA is identical to the NFA→DFA discovery construction technique we discussed above, except for the following. We would like to bundle computation-objects for *and*-states inside compound objects and use the NFA's nondeterminism for the *or*-states. In other words, all computation-objects originating from an *and*-state in the AFA's tree response should belong to the same compound-object. On the other hand, two computation-objects originating from an *or*-state should never belong to the same compound-object.

Hence, we reuse the NFA→DFA discovery technique but whenever a child compound-object is created, we scan all of its member computation-objects for an *or*-type computation-object that has next-state conflicts in A. If none exist, the process continues as in the NFA→DFA conversion. If, however, such a state exists—as in the case of the compound-object <A0,A1> in Figure 1.8c, where A1 is an *or*-type state with next state conflicts (via transitions A1→$_q$A4 and A1→$_q$A2 as in Figure 1.4a)—the behavior is as follows. Instead of adding both next states (A4 and A2) to the newly created compound-object, as was done in the NFA→DFA conversion in Figure 1.7c (resulting in the new compound-object <A4,A2>), we create two new compound-objects, one that "guesses" the use of the transition A1→$_q$A4 and the other that guesses the use of the transition A1→$_q$A2. Hence, the first new compound-object is <A4>, which uses A4, and the other is <A4, A2>, which uses A2.

The final states of *N* are all those compound objects in which *all* member computation-objects represent final states in *A*. This follows the definition of an accepting AFA trace.

The existence of this conversion technique means that in general, like NFA and ∀-FA (and ε-NFA), AFA are capable of specifying only regular requirements, because every AFA can be converted into an equivalent NFA or ∀-FA.

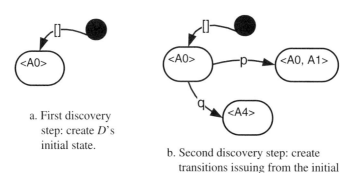

a. First discovery step: create *D*'s initial state.

b. Second discovery step: create transitions issuing from the initial state, and discover resulting states.

FIGURES 1.8a–b   Discovering an NFA that mimics the AFA of Figure 1.4a.

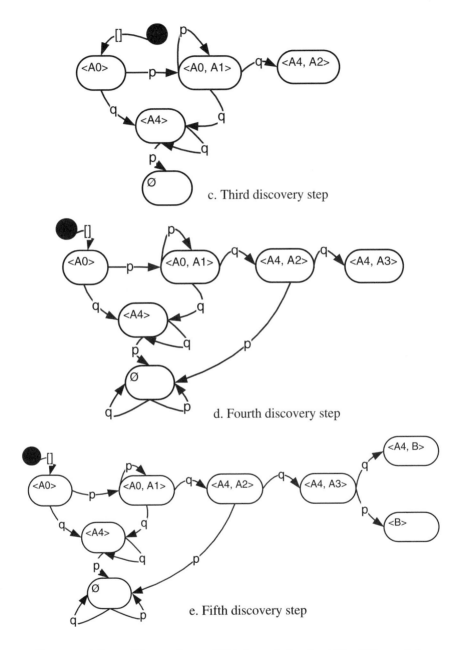

c. Third discovery step

d. Fourth discovery step

e. Fifth discovery step

FIGURES 1.8c–e   Discovering an NFA that mimics the AFA of Figure 1.4a.

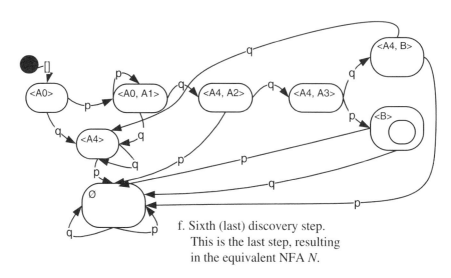

f. Sixth (last) discovery step.
This is the last step, resulting
in the equivalent NFA *N*.

FIGURE 1.8f   Discovering an NFA that mimics the AFA of Figure 1.4a.

## 1.8. Operations on Regular Requirements

Since requirements are themselves sets of scenarios, we can obviously use set operations on them, such as constructing the union or intersection of two or more requirements. Let $R_a$ and $R_b$ be regular requirements. By definition they have an equivalent DFA, which, again by definition, can be considered an NFA, ∀-FA, or AFA.

1.  $R_c = R_a \cup R_b$ is a regular requirement. This is because there exists an ε-NFA for $R_c$, shown in Figure 1.5c for $R_c$. Also, there is a statechart assertion pattern for implementing the union of two statechart assertions, as we will see in Section 3.7.

2.  $R_c = R_a \cap R_b$ is a regular requirement. That is because there exists an ε-∀-FA for $R_c$, shown in Figure 1.5d. Also, there is a statechart assertion pattern for the implementation of the intersection of two statechart assertions, just as there is for the union operator, likewise to be discussed in Section 3.7.

3. The *complement* of a requirement, R, $R_c = \Sigma^* - R$, is a regular requirement. The DFA, *D'*, that realizes $R_c$ is almost identical to the DFA, *D*, that realizes R, except that every final state in *D* is defined as nonfinal in *D'* and vice-versa. Clearly every scenario that is accepted by *D* is rejected by *D'* and vice-versa.

Note that inverting an NFA requirement isn't as trivial as inverting a DFA requirement. In addition to swapping between final and nonfinal states, we need to invert the acceptance criteria, from existential to universal, resulting in a $\forall$-FA. Alternatively, we can convert the NFA into an equivalent DFA, D, using the subset construction and then complement D as described above.

To invert an AFA requirement, first we swap between final and nonfinal states. Then we need to swap state types, so that all *and* states become *or* states and vice-versa.

## 1.9. Succinctness of FA

We have seen that DFA, NFA, $\forall$-FA,and AFA have the same descriptive power; they are all capable of describing regular requirements and only regular requirements. Nevertheless, there are significant differences in succinctness between these dialects.

The diagram in Figure 1.9 shows known upper and lower bounds for converting one type of FA into another (In Chapter 3 we will extend these succinctness results to include statecharts). For example, there is a known exponential upper bound for converting an NFA into a DFA, a bound provided by the subset construction. There is also a known lower bound for the same conversion, so that for certain requirements conversion is exponential at best.

Note that the primary reason for the exponential gap between NFA ($\forall$-FA) and AFA is that the those types of FA have no simple ability to describe the complement of a requirement.

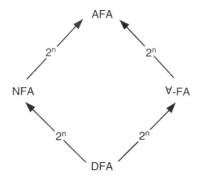

FIGURE 1.9   Upper and lower bounds for FA.

We will not prove all lower bounds associated with Figure 1.9. Nevertheless, we will discuss the exponential lower-bound blow-up associated with the NFA→DFA conversion because it provides some insight into the nature of the lower bounds. We will see that there is a requirement that can be specified with an NFA whose size is n + 1 states but for which *every equivalent* DFA has at least $2^{n-1}$ states. The NFA, N, for n = 4, in Figure 1.7 captures the following requirement described in a combination of NL and set notation:

**R1.6:** {*all scenarios from {p,q}\* that end with symbol p followed by n–1 symbols of any kind*}.

We will leave the formal lower-bound proof to a more theoretical book. However, here is a high-level explanation. Any DFA, $D$, that realizes R1.6 must somehow examine every input symbol $p$ and then start counting $n - 1$ symbols. But it can count only using states, because that is the only memory device an FA of any kind has. Therefore, it will need a distinctive state for every suffix scenario of length $n - 1$ (there are $2^{n-1}$ such scenarios) or else it can be tricked into either accepting bad scenarios or rejecting good ones, thereby violating R1.6. For example, consider $D$'s computation responses to the two input sequences $seq_3 = p.q.q.q$ and $seq_4 = p.q.p.q$. Say that the last computation-object in those two computations are C1 and C2 and that C1 and C2 are visiting the same state, $S$. We can now extend $seq_3$ and $seq_4$ by additional $q.q$, namely $seq_3 = p.q.q.q.q.q$ and $seq_4 = p.q.p.q.q.q$. If $S$ is final, $D$ is erroneously accepting $seq_3$; if it is not, $D$ is erroneously rejecting $seq_4$.

In contrast, the NFA of Figure 1.10 manages to specify R1.6 with $n + 2$ states because when it sees an input $p$ it creates two computation-objects: one assumes that this is the $p$ that is $n - 1$ symbols from the end, and the other that it is not. The first computation-object then verifies that indeed $p$ is $n - 1$ symbols from the end of the scenario.

The insight we get from this lower-bound proof is that the crux of the matter is counting. A devil's advocate would argue that with one bit-array the DFA could capture the same requirement succinctly. Indeed, Java statecharts introduced in Chapter 2 enable the use of Java variables within statecharts.

Nevertheless, in Chapter 4 we do show several requirements written as nondeterministic statecharts that are clearly more readable than their known deterministic counterparts.

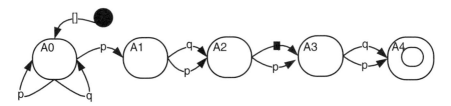

FIGURE 1.10   An NFA for R1.6 with the $n + 2$ state ($n = 3$).

## 1.10.  Specifications as Zipped Requirements

There is an interesting analogy between formal specifications (e.g., using DFA or NFA) and the decompression of compressed digital data, shown in Figure 1.11. A decompressed, or zip, file or stream is a small and compact representation of larger original data, such as text. The compressed data is not directly readable, however; to be readable a decompression (unzip) program is required.

Similarly, a formal specification, in the form of a DFA, NFA, statechart, or any other formal specification language, is a compressed representation of a requirement (a potentially infinitely large set of scenarios). Clearly, when the requirement contains an infinite number of scenarios, neither a person nor a machine can explicitly write them all. They are, therefore, represented in compressed form using a formal specification. However, to be of any practical use (especially by a computer) a requirement must be extracted from its compressed form. One such use is runtime monitoring, which we will discuss in Chapters 3 and 4. The runtime monitor is then the equivalent of a decompression program, because it extracts scenarios from the compressed representation.

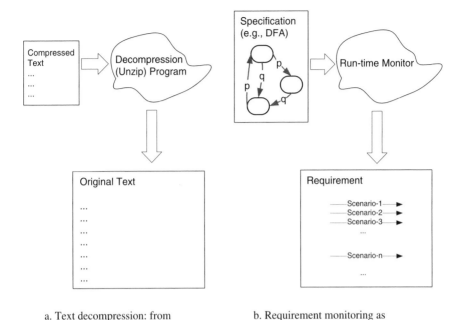

a. Text decompression: from
   compressed form to larger,
   original and readable, form.

b. Requirement monitoring as
   decompression: from compressed
   form to larger, original form.

FIGURES 1.11a–b   Text decompression as an analogy to
requirement specification and monitoring.

## 1.11.  Finite State Machines

We can think of an FA as a black box that generates a sequence of
Boolean accept/reject (1/0) output streams. For example, in response
to the input scenario $seq_1 = p.p.q.p.q$, the FDFA in Figure 1.1a generates the sequence 0.1.1.0.0.

In contrast an FSM generates output sequences using symbols
taken from an output alphabet, $\Gamma$. In other words, rather than simply
qualifying input scenarios as "good" or "bad," an FSM actually
responds to those scenarios with sequences of actions. For example,

Figure 1.12 shows an FSM controller for a simple light switch that assumes the input alphabet $\Sigma$ = {*lightOn, lightOff*} and the output alphabet $\Gamma$ = {*turnCameraOn,turnCameraOff*}. The output response of the FSM to the input scenario *lightOn.lightOn.lightOff.lightOn* is then *turnCameraOff.turnCameraOff.turnCameraOn.turnCameraOff.*

Classical FSM literature differentiates between Mealy and Moore FSMs. In a Mealy machine, output actions are specified along transitions (on the right-hand side of a virgule, or slash, "/") as in Figure 1.12a. In a Moore machine, output actions are specified inside states, as in Figure 1.12b. As we will see in Chapter 2, state-charts—which extend FSMs in many ways—combine both forms of output specification.

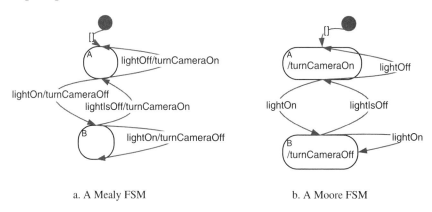

a. A Mealy FSM                    b. A Moore FSM

FIGURES 1.12a–b   FSMs.

## 1.12. Normal Form and Minimization of FA and FSMs

The normal form of an FA is a unique and minimal equivalent FA specification. DFA, FSMs and hierarchical FSMs (HFSMs, which we will discuss in Chapter 2) all have a normal form. In other words, these kinds of FA and FSMs all have a minimization procedure that yields a unique, minimal specification for every regular requirement. DFA minimization algorithms are widely available in automata and formal language literature and FSM and HFSM minimization algorithms are widely available in digital circuit design literature.

In contrast, NFA, AFA, $\forall$-FA, and statecharts (other than HFSMs) have no known normal form. Moreover, even in the cases in which a normal form does exist, it does not necessarily mean that no smaller specification can be created using an alternative specification language. For example, we saw in Section 1.6 that the best DFA specification for requirement R1.6 is at least exponentially larger than the NFA in Figure 1.10, therefore the NFA is exponentially smaller than the minimal DFA realization of R1.6.

Although minimization *per se* was considered an important aspect of FA and FSM use in the early days of computer science, and is still very important for certain hardware applications, it is mostly a theoretical issue for present-day UML users. Normal form, on the other hand, is still important for contemporary applications, like equivalence checking. If two developers create separate statechart models or assertions, it would be beneficial to be able to compare them and determine, with 100% certainty, whether or not they are equivalent.

# Chapter 2

## Statecharts

### 2.1. Transformational vs. Reactive Components

Systems, subsystems, and components are typically viewed as either transformational or reactive. Transformational components are stateless components; that is, they have no memory that persists between successive invocations of the component. Examples are a square-root computation, a fast Fourier transform (FFT) of a signal, and the compression of a voice signal. Hence, transformational systems perform a fresh computation every time they are invoked. Reactive systems, in contrast, perform an ongoing and often never-ending computation, in which each invocation uses information generated by previous invocations. There are numerous methods for representing this past memory or the information embedded in the past memory that is required for the future. *State*-based models, such as UML statecharts, are one of the most popular methods of describing the behavior of reactive components.

Finite state machines (FSMs) have been used for more than half a century to describe reactive components. Statecharts, being an extension of FSMs, continue this tradition and are the most popular language for modeling reactive components.

## 2.2. Statecharts in Brief

Before statecharts were introduced, FSMs were often considered a good representation for classroom models of reactive components, but in practice, when they were applied to larger problems, the models were cluttered and unreadable. The primary reason for this result is that FSMs are flat and sequential and therefore do not scale well when applied to large components. In fact, although programmers in the 1980s were often using top-down decomposition as a part of their software development methodology, they could not use top-down techniques to develop complex reactive components using FSMs.

David Harel introduced statecharts in the 1980s as a language for describing complex reactive systems found in contemporary avionics systems. At the time, I was his Ph.D. student investigating the first code generation technique for statecharts. In the next two decades, statecharts were adopted by the Object Modeling Technique (OMT) methodology and later by its descendant, the Unified Modeling Language (UML) standard.

In a nutshell, statecharts extend FSMs with these capabilities: events and conditions; hierarchy (state nesting); concurrence, or state orthogonality; and history states. In addition, as explained below, statecharts have built-in capabilities for describing their interaction with multiple objects in their environment.

This book considers statecharts in the context of a Java application. Rather than taking the perspective of a CASE tool, where the entire system is described within the tool, it takes a more humble perspective, where statecharts are used to describe control components within a Java system, while the rest of the system–i.e., the static, stateless, components, or data stores–are captured either by another tool (UML class and object diagrams, for example) or directly in Java.

## 2.3. A Related Tool

Throughout this book we will discuss statecharts as they relate to the StateRover™ tool. The StateRover is a statechart, Visio™-based[1], code generator and visual debug animator. It automatically generates Java code for statechart models, statechart assertions, and temporal logic assertions within statecharts and also for Java JUnit-compatible white box tests. Figure 2.1 shows the StateRover code generator and Visio working in tandem.

The repeated reference to the StateRover throughout the book is because it is the only working reference to the approach suggested in this book, and also because I personally authored the tool.

---

[1] An Eclipse GUI is currently under development.

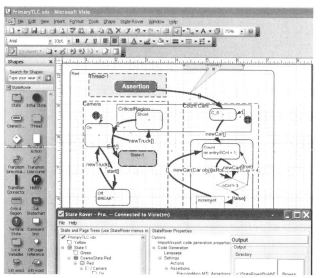

FIGURE 2.1   The StateRover consists of Visio drawing package
             and the StateRover options GUI.

## 2.4. Basic Elements of Statecharts

### 2.4.1. States and Transitions

The two basic components of statecharts, like those of FA and FSMs,
are states and transitions.

**States**

States are used for memory purposes. When a statechart visits a
state, it is a visual reflection of memorizing an artifact. This is manifested by the fact that the code generated by the StateRover tool uses
plain Java variables to implement states (the array named PS, which
stands for "present state"). The memorization performed by a statechart is about the nature of certain sequences of inputs (sequences
of events and conditions) seen so far. One reason for a statechart to

memorize particular sequences is because the controller statechart needs to respond to these sequences in a particular manner. For example, when a pump controller senses a *pumpOff* event, it memorizes that by going to a state called "Off." Obviously, the controller's response to future events depends on the fact that the pump is off. Another reason for using a state is as a helper state, which serves as a prefix for memorizing longer sequences in a subsequent state.

A state differs from a Java variable primarily in its visual appeal. In fact, UML languages as a whole are just a visual means for describing, documenting, and specifying software systems written in Java or C++–that is, they all have Java or C++ counterparts–and it is their visual appeal that is the primary reason for their prevalence.

States have potential actions. These actions range in complexity; a simple assignment to a variable, such as $x = 0$, for example, is a valid action, and so is a method call or an object instantiation. Types of allowed methods depend on the visibility of the action within the code. In the StateRover's case, we will see that the statechart model is implemented as a class within a Java project, and methods can therefore be class-level actions in this class (for instance, actions on member variables) or any method invocation of objects visible from this class. We will discuss the style of StateRover's generated code in Chapter 3.

State actions can be specified to execute when the state is entered ("on-entry" actions) or exited ("on-exit" actions) or as long as the state remains unchanged ("Do" actions).

**Transitions, Events, and Conditions**

Recall that FA transitions are annotated with symbols taken from a finite domain of discourse (alphabet) and that FSM transitions are annotated with input and output symbols, from input and output alphabets. Statechart transitions, on the other hand, are annotated with events, conditions, and actions. The general syntax of a state–chart transition is Ev[cond]/action, as shown in Figure 2.2 where reset and timeset are events, *nCarsWaiting*>0 is a condition, and *nCarsWaiting++* is an action.

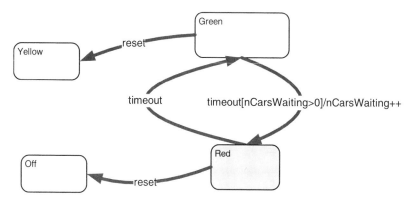

FIGURE 2.2a   The general syntax of a statechart transition is Ev[cond]/action.

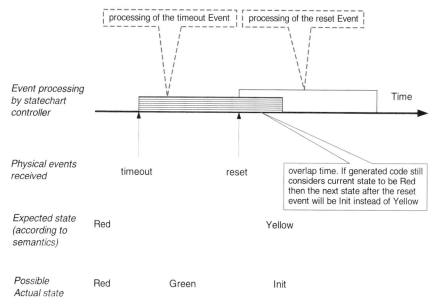

FIGURE 2.2b   Overlapped processing of events.

An event represents information sensed by the controller statechart that is available for a very short spike, or time interval, and then vanishes. It is like a single ring on your phone, which you can either respond to (take the call) or ignore, or if you were busy with other activities, perhaps even miss.

Events in themselves have no memory, or persistence, though you could create an event buffer that collects and perhaps implements more persistent events. In fact, a hardware interrupt system often works this way, where an interrupt is a hardware manifestation of an external event and the hardware maintains a memory of the events that were suspended.

A condition, as found in statechart transition guards, is any valid Boolean condition taken from within the visible scope of the application that the controller statechart, such as a condition on class member variables–for example, *nCarsWaiting* > 0 in Figures 2.2a and 2.3, where *nCarsWaiting* is a class member attribute within the statechart traffic light controller class. Hence, unlike an event, a condition persists until a point in time when the inverse condition holds.

Hence, one way to think about events is to consider them as conditions with a default value of false and no persistence. When a condition like $x > 0$ or a Boolean variable $b$ attains a true or false value, that persists until the variable $b$ or the constituent variable $x$ inside the condition $x > 0$ changes. An event, on the other hand, has no associated persistence machinery, typically because it is generated by an external entity.

The StateRover's manifestation of events uses Java methods; for example, firing a statechart event, such as *timeout* in Figure 2.2a, is implemented by calling a method, here *timeout()*. These methods are called *event handlers* [see 2.5.3]. Indeed, method calls, like events, have no persistence: when a method is called and returned from, that fact becomes forgotten history, much like the traffic light at the corner being red at 8:14:32 AM without a camera or some other recording device. To memorize the fact that the event occurred, you must draw a transition like the one from the *Green* state to the *Red* state in Figure 2.2a.

As physical entities, events consume no time: they are zero time episodes. Therefore, we can safely assume that two or more events are never received simultaneously from the physical environment of the component under design. For example, consider the events "light turns on" and "email sent." As close as they might be in time, they are never completely simultaneous. This holds even if the events are caused by a single entity, because it cannot generate them absolutely at the same time. In addition, the component receiving the events has various hardware and software layers, such as a processor's interrupt system, that process and prioritize incoming events. These layers guarantee that when events are received in close temporal proximity, then one event is always treated before the other.

It is desirable to make the model as close as possible to the phenomena modeled—i.e., to assume that the events are never simultaneous. In the implementation, however, the software actually consumes time when processing those events. Hence, for the statechart of Figure 2.2a, a "reset" event might be received while the software is processing a "timeout" event while in state Red, as illustrated in Figure 2.2b. In this case, unless the code generator takes special precautions, the software could consider that "Init" is the next state although it should semantically be "Yellow." A similar situation could occur if *reset* and *timeout* events are generated by (called from) concurrent threads within a single Java or C++ application. A simple extension to the StateRover's generated code, using synchronized event buffers, provides a *robust* code generator that can handle such event preemption. Robust code generation is discussed in Section 2.5.5.

Like state actions, transition actions are actions performed when a transition is traversed. Like conditions, they are statements that refer to visible methods and variables from the scope of the application, such as the increment action *nCarsWaiting++*.

Statechart events and conditions are potentially generated by multiple concurrent external entities or objects. Similarly, statechart actions might trigger, or feed data to, several external objects. The input-output behavior of statecharts is therefore like that of a multitape FSM. However, as we will see, statechart orthogonality is a better way of internally representing how the controller interacts with multiple objects in its environment than the FSMs purely flat, sequential state-based approach.

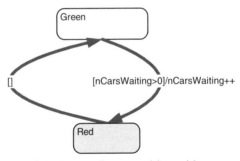

FIGURE 2.3   A statechart transition with no events.

Statechart transitions with no events imply the use of a hidden "clock tick." Hence, for example, the transition from Green to Red in Figure 2.3 is presumed to take place in the next cycle, or clock tick. The clock tick is then defined outside the statechart traffic light controller. A statechart in which transitions have no events is often called *procedural*, in contrast to the event-driven type described above. In the sequel, we will show the StateRover approach to realizing event-driven statecharts, procedural statecharts, and mixed statecharts—i.e., statecharts with a combination of event-driven and procedural transitions, where we use flowcharts inside statecharts to capture both types of control flow within one diagram.

**Internal vs. External Events and Event Preemption**

Developers are occasionally concerned about the origin of an event–
that is, whether it is internal or external to the package or application
under development. From the standpoint of the controller statechart,
what really matters is whether events are generated inside the controller or by an external entity. For example, the timeout event of
Figure 2.2 could be generated as an action on some transition or in
some state. In this case, the statechart would be sending itself mes-
sages, using the timeout event as a message. There are two problems
with such a design:

1. *Stability.* You could potentially create an unstable design
   as a result of an infinite loop of transitions where one transition induces the traversal of another by generating its
   event. For example, if both timeout transitions in Figure
   2.2 also generate a timeout event, the controller statechart
   could potentially traverse, in a single cycle, the infinite loop
   Green→Red→Green→Red. . . .

2. *Preemption.* Consider a transition, labeled *Ev1*[]/*Ev2*,
   from state *A* to state *B* and another transition, labeled
   *Ev2*[], from state *A* to state *C*. Depending on the code
   generator, the action *Ev2* could fire before the state bookkeeping registered a move out of state *A*. In this case, the
   resulting state would be *C* instead of *B*. This would be a
   serious error because the model's semantics would differ from the actual behavior of the generated code. The
   StateRover tool addresses this issue in two ways: it checks
   for runtime event handler preemption, and it lets you drive
   the code generator to catch all instances of potential event
   handler preemption.

**State Hierarchy**

Hierarchy, or state nesting, is the ability to nest states—i.e., to draw states within states. Figure 2.4a illustrates state hierarchy with the StateRover tool. Hierarchy is usually represented using two drawing techniques: *explicit nesting* and *coarse states*. These are merely different visualization techniques with identical semantics. Explicit nesting is when we explicitly draw a lower-level state inside a higher-level state, or *superstate*, such as *Green* or *Red* inside *Active* in Figure 2.4a. A coarse state is a state whose contents are drawn on a separate drawing canvas (e.g., a new Visio page), such as the state *Red* shown in Figure 2.4a.

Hierarchy serves four purposes:

1. *State refinement, mostly for purposes of top-down design.* For example, suppose that we want to refine the Red state in Figure 2.2 to consist of substates. Within the StateRover tool, we can make this refinement either using explicit nesting, inside Red's state circle, or by defining Red as a coarse state and drawing its substates on a separate "page."

2. *Reduction of transition clutter and transition state dependence.* Consider Figure 2.4. Figures 2.4a and 2.4c are semantically equivalent, but Figure 2.4a is more readable. Note how, in Figure 2.4a, the "reset" transition from the Active superstate to the Yellow state has the same meaning as the two direct transitions: Green→Yellow and Red→Yellow. At first glance, such hierarchical transitions seem to offer a rather trivial improvement over direct FA-or FSM-like transitions. However, note that if, at a later time, a different modeler undertakes further state refinement and adds additional substates within the Red superstate, the high-level "reset"

transition to Yellow eliminates the need for the new modeler to be aware that those new states must be connected via a "reset" transition to Yellow to achieve the desired behavior.

3. *Orthogonality*. Superstates are place holders for orthogonal activities, described below.

4. *Shared actions*. Consider Figure 2.4a again. A Do in the Red state will execute as long as the traffic light controller is in any of the Red state's substates.

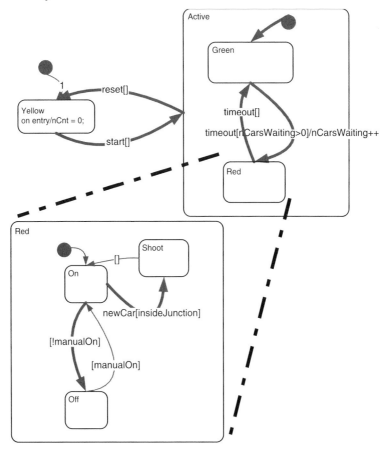

FIGURE 2.4a   State hierarchy (nesting).

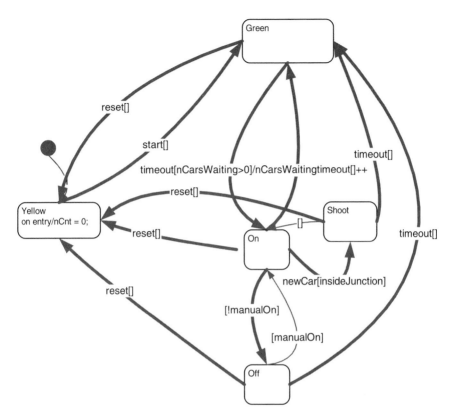

b. The equivalent diagram without state nesting

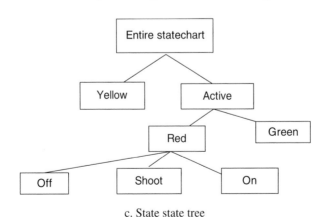

c. State state tree

FIGURES 2.4b–c   About State hierarchy.

Like FA and FSMs, a statechart has an initial state. In fact, every superstate, including the entire statechart as the "supreme" superstate, has a default, or initial, state. In Figure 2.4, for example, *Active* has *Green* as its default state, and *Red* has *On* as its default state. Thus, when the transition *Yellow→Active* occurs, *Active* will start operating in the *Green* state.

Hierarchy induces the statechart *state tree*, as illustrated in Figure 2.4c, where inner states are drawn as descendants of their immediate superstates. We will extend this state tree to an and-or tree once we introduce the concept of orthogonality, below. All states that are leaves of the statechart state tree are called *atomic* states; all other states are called *superstates*.

### Races, nondeterminism, and transition priorities

A modeling statechart must be deterministic, unlike an assertion statechart, discussed in Chapter 3, which can be nondeterministic. A race exists when, given an input event, a conflict occurs between the resulting next states. Figure 2.5a illustrates two races: one following the *timeout* event while the statechart is in the *Green* state; the other following the *start* event while it is in the *Yellow* state and *nCarsWaiting* = 0. In the first case, the race, or conflict, is between *Red* and *Yellow* being the next state, because only one of them can be next. In the second case, the race is between the *Yellow* state and the *Green* state. However, it is commonly accepted that when races exist, the transition drawn higher in the statechart hierarchy is assigned a higher priority and is the one to be executed; therefore, the "timeout" race is resolved, the next state being *Yellow*. The second race, however, is a real race. The StateRover tool takes the following actions to detect, warn of, and even eliminate races:

- It optionally issues code generation warnings for all potential races. In this case, we would be notified of the potential conflict induced by the start event and could resolve it using, for example, an improved statechart, like the one in Figure 2.5b.

- It optionally generates code that detects races during run time. It then chooses one of the conflicting transitions over the other and executes a custom action (like print, error reporting, or alternate recovery method).

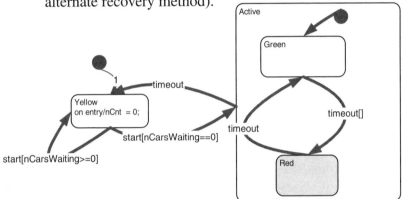

FIGURE 2.5a   Potential races.

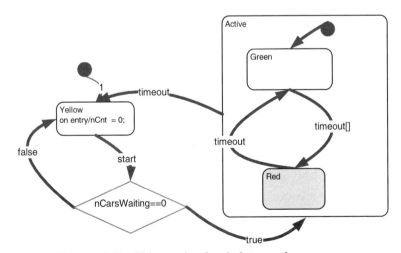

FIGURE 2.5b   Using a visual switch to resolve a race.

Up to this point, we get what are known as hierarchical state machines, or HFSMs. An important feature that distinguishes statecharts from HFSMs is state *orthogonality*, better known as *concurrence*.

**The Semantics of HFSMs**

In this book, I use a form of semantics, or formal meaning, known as *interlingua-based* semantics, in which the semantics of HFSMs are given in terms of another, well-defined language. We therefore provide a simple translation from HFSMs to multitape FSMs. The translation, shown in Figure 2.4 is simple. First, we modify every transition that has a superstate, $S$, as its target state to have an atomic state as its target, as in Figure 2.4a, where we would modify the transition *Yellow→Active* with the transition *Yellow→Green* while preserving the transition event, condition guard, and action. Next, we substitute every transition that has superstate $S$ as its source state with explicit transitions, one from every atomic state under $S$. The substitute transitions have the same event, guard, and action as the original. Figure 2.4b is the equivalent FSM to the HFSM of Figure 2.4a.

**Actions**

Statechart actions can be generated at three places, as shown in Figure 2.6a: along transitions, such as the *syncFromCounter()* action; on entry to a state, such as the *enteredRed()* action; and on exit from a state, such as the *exitedRed()* action. In addition, there are Do actions, such as the Do action in the *Red* state of Figure 2.7a, which is executed as long as *Red* is the present state, excluding the cycles in which *Red* is entered and exited. Such a Do action will execute even if the statechart changes states while in the *Red* state. Hence, in Figure 2.7a, upon transitioning from *On* to *Off*, the traffic light controller (TLC) fires the *Off* state's on-entry action and the *Red* state's

Do action. The order of execution of these actions depends on the tool; with the StateRover, you can specify the order.

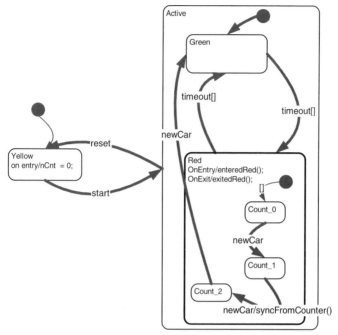

FIGURE 2.6a  An HFSM. HSFM#1 for the flat
TLC implementation (Section 2.4.2).

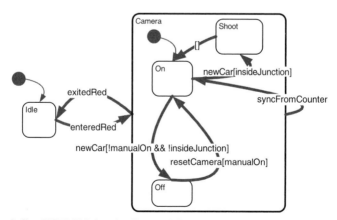

FIGURE 2.6b  HSFM#2 for the flat TLC implementation (Section 2.4.2).

This example illustrates a situation in which multiple actions are executed in the same cycle—that is, when an external event fires. Another example, related to Section 2.4.2 below is an on-entry action in one thread firing in the same cycle in which a transition action fires in another thread.

With the StateRover's vanilla code generator, all actions scheduled to execute in a given cycle are scheduled using a fixed, non-interleaving schedule, that is, one after the other. When they all complete, the cycle is considered to have ended. The concurrent-actions code generator creates code in a more sophisticated way. (The code generators are described below.)

## 2.4.2. Concurrence, or Orthogonality

*Concurrence*, or rather *orthogonality*, is illustrated in Figure 2.7a. When in the *Red* state, the statechart traffic light controller performs two FSM-like computations, one controlling a camera and the other counting cars. For the most part, these activities are independent of, or orthogonal to, each other and are therefore drawn as two *threads*, depicted by dashed boxes. Each statechart thread can be conceptually regarded as a statechart in its own right. In general, statechart threads can have hierarchy (nesting) and internal orthogonality. It is important to distinguish between *statechart threads* and OS-level or Java threads; these are different concepts, as discussed in detail below.

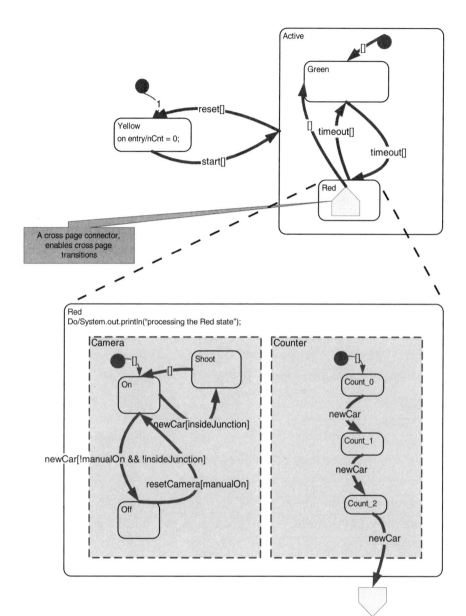

FIGURE 2.7a     Statechart with hierarchy and concurrence.

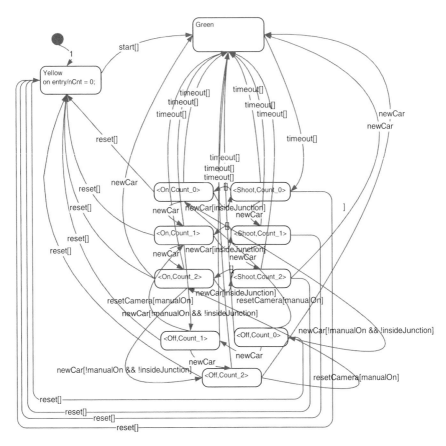

FIGURE 2.7b   The equivalent of FSM.

Orthogonal threads behave like a conceptual Cartesian product machine, as illustrated in Figure 2.7b, which is the equivalent Cartesian product FSM for the statechart in Figure 2.7a. Assume that the statechart enters the *Red* state after a *timeout* event. It then actually enters a pair of states: *On* in the *camera* thread and *Count_0* in the *counter* thread. If the *newCar* event now occurs, the *camera* thread transitions either to the *Shoot* state or to the *Off* state, depending on the associated transition guard conditions–for example, to *Shoot* if the *insideJunction* Boolean condition is true.

Note that if all transition guards are false, the *camera* thread stays in the *On* state. Upon the same *newCar* event, the *counter* thread transitions to the *Count_1* state. The overall state of the statechart is called a *state configuration*, denoted in this case as <On,Count_1>. Indeed, this seems like concurrent behavior. Note, however, that there are many ways to implement this behavior. In fact, nothing in the behavior as described so far says anything about hardware or OS-level concurrence; the only commitment made is that from the state configuration <On,Count_0>, upon the *newCar* event and if the *insideJunction* Boolean condition is true, the next state configuration in the next cycle will be <Shoot, Count_1>. The Cartesian product machine is used for the interlingua semantics for statecharts.

An event can trigger transitions in all active threads, in some active threads, or in none. For example, if the *start* event is sensed, while the traffic light controller statechart is in the <On, Count_0> pair of states within *Red*, no thread responds. Nevertheless, the Do action of the *Red* state will execute because the statechart is still in the *Red* state.

Statechart orthogonality can be partial—that is, there can be some dependence between the orthogonal activities. For example, once the *Counter* thread in Figure 2.7a detects three cars, it changes the state within *Active* to *Green*, thereby aborting all computations inside *Red*, including the *Camera* thread. In other words, that *Camera* thread is not completely orthogonal to the *Counter* thread.

*Visual synchronization* is used to visually capture the way some threads synchronizes other threads. For example, in Figure 2.8—a modified version of the *Red* state of Figure 2.7—when the second *newCar* event is detected, it not only induces a state change within the *Counter* thread, it also forces the *Camera* thread to change to the *On* state from whatever state it was in (e.g., *Off*).

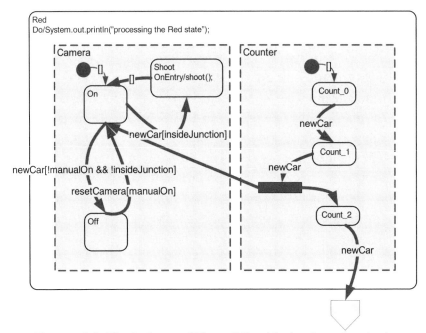

FIGURE 2.8   The *Red* state of Figure 2.7a with visual synchronization.

As previously mentioned, statechart orthogonality is often called *concurrence*. This term is misleading, though, because users often expect to see some multithreading in the resulting code, which is not necessarily the case. Consider, for example, Figure 2.7a. The StateRover's generated code for each *newCar* transition is a block of at most ten lines of simple Java code. There is no apparent benefit in performing these blocks concurrently rather than using a predefined schedule (such as the *Counter* thread's transition first, then the *Camera* thread's transition). In fact, the decision to execute these transitions sequentially or concurrently (whether time-shared on a single CPU or using multiple CPUs) has little to do with the reason for having two separate statechart threads in the model. Rather, the orthogonal threads in the statechart model exist because of the nature of the controller, which controls two orthogonal activities: camera management and counting cars.

State orthogonality is a statechart feature that is often used to represent the controller's reaction to, and control of, independent agents in its environment, as shown in Figure 2.9, where the the traffic light controller statechart controls, and reacts to, five different entities. In this respect, there is concurrence—i.e., the external agents or entities are concurrent. The TLC statechart needs to perform a logical separation of the two associated control behaviors, one for the camera and the other for the counter, but not necessarily in a truly concurrent manner. Instead, it can perform those control decision and action computations serially. If these computations are relatively short, there is no need to execute them concurrently.

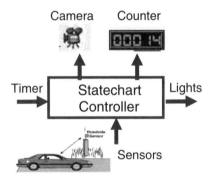

FIGURE 2.9   A statechart controller controlling
and reacting to five entities in its environment.

One reason for having a multithreaded implementation of the controller statechart occurs when the statechart threads issue time-consuming actions, thereby inducing a delay on the other statechart threads. For example, say that in Figure 2.7a the *newCar* event fires while the controller is in state configuration <On, Count_0>, and say that the transition in the *Camera* thread fires first. If this transition induces an action that performs a relatively long operation (a transition action or an on-entry action in the next state, for example), such as a network access, the second transition, from Count_0 to Count_1, will be delayed until the action completes. In this case, there is sufficient reason to implement the transitions in a multi-threaded manner.

The StateRover's solution to this issue is simple: its "plain vanilla" code generator generates purely sequential code with a fixed schedule for orthogonal transitions. A special "multithreaded," or concurrent, code generator implements the statechart as a collection of threads: a main thread for the controller statechart excluding the actions and separate threads for actions. In fact, you can select particular transitions whose actions will be segregated in separate threads.

The bottom line is that the term statechart *concurrence* relates not to the internal implementation of statechart orthogonality, but more to the environment in which it is operating. Indeed, the statechart TLC is *controlling concurrent entities in its environment*, namely, the Camera and Vehicle counting.

A common misconception is that statechart orthogonality can be easily substituted with a plurality of concurrent HFSMs, a design often called *flat* because concurrence is not nested within the statechart hierarchy but present only at the highest level of the controller. In fact, some well-known UML tools do not support statechart concurrence under the pretense that you can easily describe the same behavior using concurrent HFSMs. Although such a conversion can be performed, it places significant additional burden on the modeler to model the desired behavior correctly. For example, Figure 2.6 illustrates the "flat" concurrent HFSM alternative to the TLC statechart in Figures 2.7a and 2.8. This solution uses two HFSMs, shown in Figure 2.6a and 2.6b. Note here how the concurrent HFSM modeler had to explicitly add the following events and states:

- HFSM#1 generates three special events, for use by HFSM#2 only, namely, *enteredRed, exitedRed and syncFromCounter*.

- HFSM#2 has an additional "Idle" state, plus three new transitions, for the events *enteredRed, exitedRed and syncFromCounter*.

Indeed, the deeper the nesting of the orthogonal threads within the statechart, and the more dependency between those threads that exists (in the form of visual synchronization), the greater the burden placed on the modeler when he or she needs to convert the statechart into concurrent HFSMs.

### Multisource/Multitarget transitions

So far we have seen transitions with a single source and a single target. A statechart with orthogonal threads can contain transitions with multiple, orthogonal sources or targets, or both. Such a transition is often called a "complex transition." An example is the complex transition from the *Count_1* state to the orthogonal pair of states *On*, *Count2* in Figure 2.8. A complex transition uses a transition connector (the tick line in the figure) to assemble multiple sources or multiple target states (or both). Note that all source states must be pair-wise orthogonal; likewise for all target states.

## 2.4.3. History

History connectors provide a mechanism for memorizing, and returning to, previously visited states within a given superstate. The most commonly used history connector is the *deep history*, or H*, connector. Consider the simplified TLC statechart in Figure 2.10a. Figure 2.10b, c, d and e demonstrate the behavior of the history connector within this TLC, as follows (thick black transitions and states represent a transition firing in a given cycle and the resulting state): say that at some point the *Red* state is entered (Figure 2.10b), followed by a certain number of cycles spent in *Red* (Figure 2.10c). Say the timeout event then occurs while the TLC is in the *Count_1* state within *Red*, resulting in a transition to a state outside of *Red* (Figure 2.10d). Upon the first subsequent return to *Red* via the H* connector, the actual next state will be *Count_1* (Figure 2.10e) because it was the most recent state visited within *Red* (Figure 2.10c).

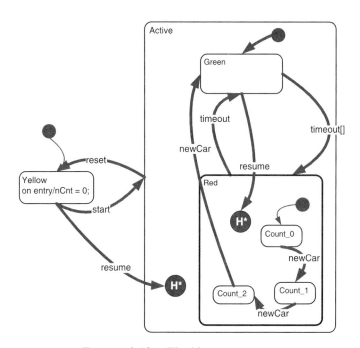

FIGURE 2.10a   The history connector.

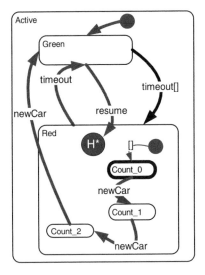

b.

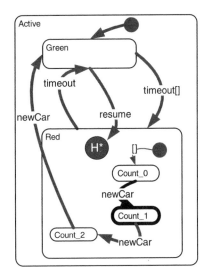

c.

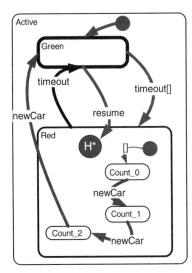

d.

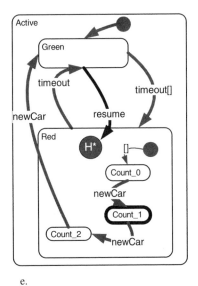

e.

FIGURES 2.10b–e.

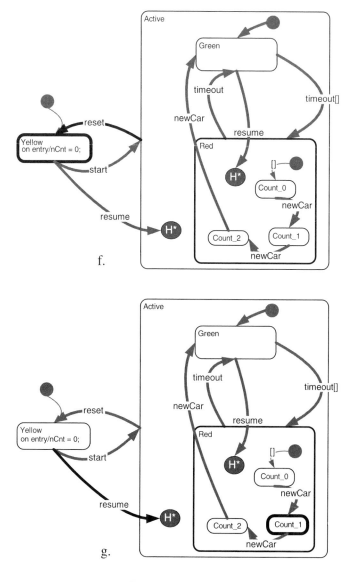

FIGURES 2.10f–g.

Note how the H* connector implements "deep" history: it memorizes that latest state visited, including all substates visited.

To illustrate that the H* connector implements deep history, let's continue the scenario of Figure 2.10. Say a *reset* event fires. (Figure 2.10f), resulting in the TLC statechart moving to the *Yellow* state. When the *resume* event fires the transition to the H* connector within the *Active* state, the TLC statechart returns to the most recently visited state within *Active*, *Red*, and within *Red* to its most recently visited state, *Count_1* (Figure 2.10g). In contrast, "shallow" history, depicted using the H connector, uses history information only in the level of hierarchy of the H connector, and thereafter uses default states information. Such an interpretation for Figure 2.10f and Figure 2.10g would result in *Count_0* as the next state, because the statechart would return to *Red* using history information but would return to the default state within *Red*.

**Return vs. History**

History in general and deep history in particular provide a mechanism for memorizing the last state visited within a certain superstate context. It is therefore tempting to model a return statement, of the kind supported by common programming languages and microprocessors, using the history connector.

Consider Java method call. A return statement within this method actually returns the program control to the statement *following* the call to that method, not to the statement that called it, as expected from the behavior of the history connector. Similarly, a return command on a modern microprocessor sets the processor's program counter (PC) to the address *following* the address of the method of function call.

The difference in the behavior between statechart history and the return construct is that a return statement is defined in an environment where the notion of the *next-command* is well defined, such as the next-command in a Java program is the one written below the current command. In other words, the return construct works in an environment where there is a natural linear progression of a computation, one that can be drawn as a state diagram with a linear chain of states. A statechart in general, on the other hand, is not necessarily linear.

## 2.5. Code Generation and Scheduling

Code generation for flat FSMs is rather simple. The FSM's state space is coded as integers, namely every state is identified by an integer symbol. A variable named PS is used to memorize the currently visited state. The transitions of the FSM are then coded as a switch statement, as follows:

```
switch (PS) {

    case Green:

        /* if statement for transition #1 from state
        Green to state Red*/

        if (guard) {

          PS = Red;

        }

        … // if statement per transition out going Green

      break;


      case Red:
```

```
    ... // if statements for transitions outgoing
        state Red

  break;

    ...

}
```

Code generation for HFSMs is also rather simple. One method, for example, is to convert the HFSM to an FSM using the *inter lingua* semantics described in Section 2.4.1, and to implement the resulting FSM as described above.

## 2.5.1. Hierarchical (Horizontal) Code Generation

David Harel, the inventor of statecharts, and the author invented a statechart code generation technique I call a *hierarchical* method. With this method, a statechart is implemented as a hierarchical structure of communicating HFSMs.

Figure 2.11b illustrates this hierarchical structure when applied to the TLC of Figure 2.11a. Each HFSM in the structure represents a superstate in the statechart. Each HFSM, except for the topmost one, has a new *Idle* state, indicating that the corresponding statechart superstate is not currently being visited. In addition, each HFSM sends and receives special events to and from connected HFSMs. For example, the HFSM for the *Counter* thread sends the *counterStopRed()* event to its parent HFSM, indicating that the parent (the HFSM for *Red*) must change states because of a decision in the *Counter* HFSM. The parent indeed changes states to its Idle state and informs its own parent (the top HFSM) to change states to *Green* using the *redSaysGreen()* event.

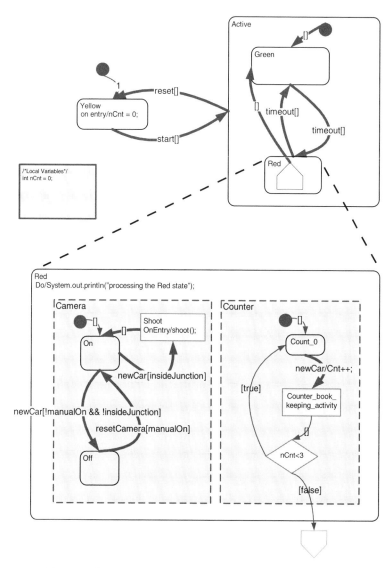

FIGURE 2.11a   The Red state of the TLC statechart with
embedded flowchart (explained in Section 2.7).

This code generation technique is often referred to as a *horizon-*
*tal* technique because the state tree is sliced horizontally. HFSMs
are then created within each horizontal slice.

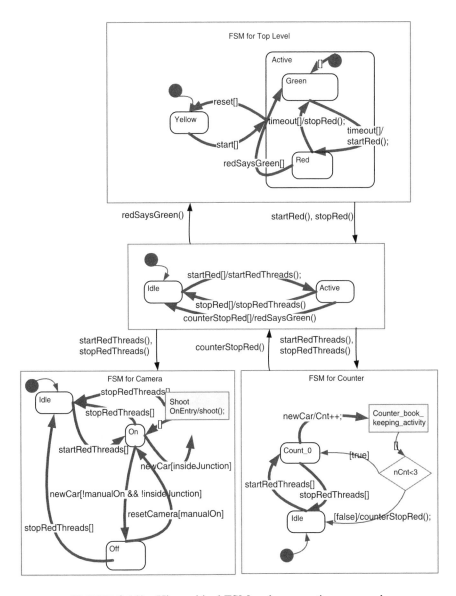

FIGURE 2.11b   Hierarchical FSM code generation approach.

Some commercial statechart code generators use this technique. The StateRover, on the other hand, uses a different technique, named the *flat* implementation. The primary reason for the flat implementation is that it results in simple and readable code, which is ultimately what is needed in order to take ownership and responsibility for the resulting code when building safety-critical systems.

## 2.5.2. Vertical Code Generation

The vertical method was illustrated in Figure 2.6 for the TLC statechart in Figure 2.8. This method constructs an HFSM for a vertical slice of the statechart that does not contain orthogonality. For example, as described in Section 2.4.2, the HFSM in Figure 2.6a contain the entire statechart, including all nested states, except for the *Counter* thread. A separate HFSM is then needed for the *Counter* thread.

## 2.5.3. The StateRover Vanilla Code Generation Style

The StateRover's code generator generates a class per statechart model, as illustrated in Figure 2.12a for the TLC statechart of Figure 2.11a. This statechart class is a convenient level of encapsulation for a controller statechart that lives within a heterogeneous system of Java or C++ objects created by various tools or perhaps hand-coded. The class can then be dynamically instantiated according to the needs of the system.

```
class TLC {


TLC() {

    eventTRreset();

} // constructor
```

```
/* method called when the statechart object boots
up*/

 void eventTRreset() {

  ...

 } // eventTRreset()


 /* Event handler for event:timeout()*/

int timeout(){

...

} //timeout()


 /* Event handler for event:newCar()*/

int newCar(){

...

} //newCar()


   /* Event handler for event:start()*/

   int start(){

 ...

   } //start()


   /* Event handler for event:reset()*/

   int reset(){

 ...
```

```
  } //reset()


} // class TLC
```

FIGURE 2.12a    Structure of the code generated
             for the TLC statechart in Figure 2.11a.

```
/* Event handler for event:newCar()*/

int newCar(){

...


  /* if-block for transition On->Shoot*/

  ...


  /* if-block for transition On->Off*/

  ...


  /* if-block for transition On->Counter_book_keep-
ing_activity*/

  ...


/*final book-keeping activity for event handler*/

...
```

FIGURE 2.12b    Structure of the *newCar* event handler.

The following sample illustrates user code for a program that instantiates and boots up the TLC statechart object whose class is listed in Figure 2.12a and then fires two events:

```
TLC myTLC = new TLC(); // instantiating and booting-up
                       // a statechart object
myTLC.start();
myTLC.timeout();
```

The controller class consists primarily of event handlers, one per transition event. For example, a statechart with three transitions annotated *Ev1*[x>0], *Ev2*, and *Ev1*[x<=0] will have two event handlers, one for *Ev1*, with code for two transitions, and the other for *Ev2* with implementation code for a single transition. The *newCar* event handler of Figure 2.12a is shown in Figure 2.12b. All transitions with no annotating event are bundled together in a single event-handler called *eventTRfire*. Although *eventTRfire* is an ordinary event handler, it represents transitions with no explicit event and is therefore expected to be fired by a scheduler of some sort.

Each event handler contains code for one or more transitions. This code is essentially a simple condition *if* block, such as the following, which implements the transition from the *On* state to the *Shoot* state in Figure 2.7:

```
try {
      if (PS[2] == ST_On) {
      if (insideJunction) {
          NS[2] = ST_Shoot;
          // actions (on exit action for source
          // state,
          // on entry action for target state, and
          // transition action)
```

```
            }

        }

    }

} catch {…}
```

Note the try-catch wrapping. You define this exception handling on a transition-by-transition basis via the StateRover GUI.

This simple code generation structure yields code that is easy to read and efficient. This basic code generator is called the *vanilla* code generator.

One downside of this event-handler-based style of code is that a caller system must be aware of the statechart event name space. For example, say an external system is built to call a *foo* method on a plurality of controller statechart objects with the expectation that if a controller has no *foo* method it will simply ignore the call. Calling foo() directly on Figure 2.12a's TLC code will obviously result in an error. For this reason, StateRover also generates a central *event dispatcher,* named *execTReventDispatcher.* It is used as a centralized location for invoking statechart event handlers; calling the dispatcher with a nonexistent event name results in the statechart executing nothing and immediately returning control to the caller. The dispatcher is also used by a primary statechart when passing methods down to a sub-statechart, as described below.

Statechart orthogonality is implemented by the vanilla code generator using a fixed schedule created during code generation. For example, in Figure 2.7a, five *newCar* transitions, two in the *Camera* thread and three in the *Counter* thread, will be realized as five *if* blocks within the *newCar* event handler. The order of these *if* blocks induces a fixed firing schedule for corresponding transitions.

## 2.5.4. StateRover Code Generation for Concurrent Actions

Consider Figure 2.13. It has two statechart threads with on-entry, on-exit, and transition actions. The vanilla code generator generates two *if*-blocks in the *newCar* event handler, one for each transition, and the order of the blocks induces a fixed schedule for the orthogonal transitions. Suppose the order is *Counter* thread (right thread) transition fires first, then the *Camera* thread (left thread) transition fires. The order of execution of actions is then:

mExit1();  mTrAction1();  mEntry1();  mExit2();  mTrAction2(); mEntry2();

A power user might need to fire these orthogonal transitions in a more concurrent manner. For example, suppose mExit1() is time consuming. This will delay the start of all actions associated with the *Camera* thread, which might be an issue if the *Camera* thread actions need to execute soon. The StateRover's code generator optionally generates code for such a user, resulting in the following order of execution for the actions in the newCar event handler:

mExit1();  mTrAction1();  mEntry1();  /* a single Java thread, sequentially executing actions associated with transition in the *Counter* thread */

------ concurrent to:

mExit2();  mTrAction2();  mEntry2();/* a single Java thread, sequentially executing actions associated with transition in the *Camera* thread */

Note in the figure the use of gray lines with hollow arrowheads to visually represent concurrent transitions, those whose actions are code generated in a separate thread. Hence, in general, the concurrent code generator induces m + n + 1 threads for handler newCar(), as follows:

- *n* Java "action" threads for *n* concurrent transitions labeled with the event *newCar*.

- *m* Java "action" threads for *m Do* actions that are enabled in the current state configuration.

- The main Java thread, which is the thread responsible for the overall statechart control state change bookkeeping and also contains the code for actions on nonconcurrent transitions labeled with event *newCar*.

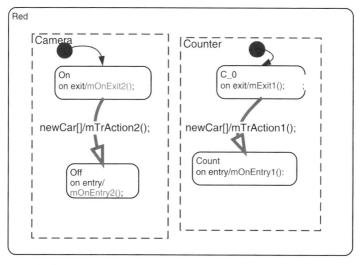

FIGURE 2.13    A statechart with concurrent transitions. Open arrowheads mean the actions associated with the transitions are concurrent, not interleaved.

Note that the vanilla code generator is a subset of the concurrent code generator in which all actions are executed in the main (and only) Java thread.

The concurrent code generator enables concurrent action threads to be executed either synchronously or asynchronously. With synchronous execution, the main thread, which, remember, is responsible for the overall statechart control state bookkeeping, will wait until all action threads complete—i.e., until all actions have executed. As its name indicates, the asynchronous mode of operation does not impose such synchronization. The setting of the synchronous or asynchronous mode of operation is performed in run time via a special API method that is generated by the StateRover code generator.

A *synchronizing state* is a state that forces synchronization even if the current mode is asynchronous. It is visually depicted using a thick black oval.

## 2.5.5.  Robust Code Generation

As illustrated in Figure 2.2b for the statechart of Figure 2.2a, when events are received in sufficiently close proximity there is a possibility that event processing by the vanilla code generator will result in erroneous behavior, namely in behavior that does not match the expected behavior as defined by the statechart semantics where events are assumed to consume zero time.

In practice, since the processing of transitions is done in an if-block of about ten lines of Java or C++ code, this situation will not occur very often. Nevertheless, a trustworthy code generator must be capable of treating such incidences.

The robust code generator does so using one of two methods. With the first method all event handlers are declared as synchronized, thereby prohibiting preemption by some other event handler.

With the second method, the statechart controller uses a Java synchronized event queue that performs event preprocessing for it. All external events are sent to this buffer first. The buffer queues them and submits them one by one to the statechart controller for processing. It releases an event for processing only when all previously received events have been processed by the statechart controller. Using a smart multithreaded implementation that is outside the scope of this book, the buffered method results in somewhat more efficient code than the first method.

## 2.6. Event-Driven Statecharts, Procedural Statecharts, and Mixed Flowcharts and Statecharts

As explained earlier, a statechart transition is annotated with an event. A statechart in which all transitions have annotating events is called an *event-driven* statechart. In contrast, a *procedural* statechart is a statechart with no events annotating transitions. Statecharts might contain a mix of both types of transitions, some annotated with events and some not (denoted as *null-event* transitions). Null-event transitions have an implicit clock event that fires the transition, an event that is supposed to be produced internally by the system under design by calling the generated *eventTRfire* event handler.

An event-driven statechart is built to expect and respond to external events. An external system fires a statechart event by calling its event handler. As described earlier, if you try to make event handler calls from within the controller statechart, it might result in event preemption and consequently erroneous behavior. Nevertheless, statechart designers often want to design a system that responds to external events but then, under certain circumstances, computes a workflow sequence of actions, one by one. Figure 2.14a shows a naïve statechart that captures such behavior. The statechart is naïve because it ignores the scheduling issue: events *reset*, *begin* and *end* are external, whereas the null-event transitions are fired from a scheduler via the *eventTRfire* method. But how will the scheduler know that it is time to call the *eventTRfire* method? And how will it know when not to call the *eventTRfire* method? Figure 2.14b shows a superior alternative, using flowcharts within statecharts.

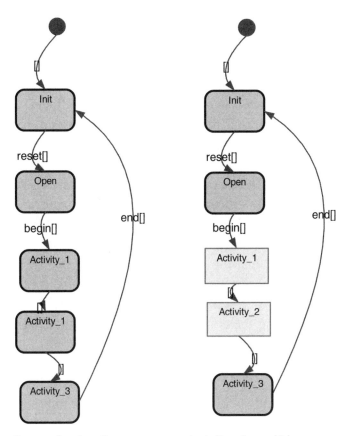

a. A naive diagram that describes          b. A flowchart within
   workflow within a statechart.              a statechart.

FIGURES 2.14a–b.

## 2.7. Flowcharts inside Statecharts: Workflow within Event-Driven Controllers

This section describes how to use flowcharts inside statecharts for the purpose of describing hybrid procedural and event-driven controllers. We use flowcharts to represent procedural computations discussed in Chapter 1, which are typically used to describe workflow.

Flowcharts are typically used to describe a process–that is, an ordering of actions, such as "first do action-1, then action-2, then if x > 0 do action-3 else do action-4." Usually, they depend only on conditions that are available at the time the flowchart is invoked. Therefore, a flowchart can execute from start to end as fast as the platform computer allows. This contrasts with an event-driven statechart, which waits in certain states for specific, external events. The statechart in Figure 2.14b behaves in the normal, event-driven manner until the *begin* event is sensed while it is in the *Open* state. Then it flows through the flowchart part visiting the flowchart *action boxes* and *visual-switch* decision polygons on the way, until it reaches a state, which happens to be *Init*. The developer of Figure 2.14a probably had the same intention but created a purely event-driven statechart instead. Consequently, when the controller reaches the *Activity_1* state it is expected to immediately transition to state *Activity_2* using the empty event; in practice, however, some other part of the application needs to generate *eventTRfire* that fires that transition. Since *eventTRfire* is not external, the developer will need to construct a scheduler that does so internally. This is a complex and roundabout way of achieving the same as Figure 2.14b.

Figure 2.11a likewise contains an embedded flowchart. When the *Counter* thread senses the *newCar* event, it flows through the flowcharts *Counter_book_keeping_activity* and the subsequent visual switch and then rests either in the *Count_0* state or in the *Green* state. Likewise, the *Camera* thread flows through the Shoot activity box when it decides to take a shot.

The StateRover therefore considers two types of states: *resting states*, which are the normal rounded box states, and *nonresting states,* which are the flowchart activity boxes and visual switch polygons. A statechart execution cycle is the consequence of firing an event (either external or *eventTRfire*). If no transition or *Do* action fires, the statechart might do nothing during an execution cycle. Alternatively, the statechart might fire one or more transitions (it might be more than one if two or more concurrent transitions fire), with each such transition-firing resulting in a change of state, from one resting state to another, possibly going through several nonresting states on the way. Hence, in Figure 2.14b, when the present state is *Open* and event *begin* is sensed, then the statechart flows to the *Activity_3* resting state in a single cycle while executing the actions in *Activity_1* and *Activity_2* activity boxes on the way and *Activity_3* in the end.

As with any flowchart, you can erroneously create a flowchart with an infinite loop. The StateRover code generator therefore generates an optional *watchdog* that checks a user-defined loop count; once a flowchart exceeds the watchdog limit a custom action is executed.

## 2.8. Nonstandard Elements of Statecharts

The StateRover enables the following useful, but nonstandard, statechart elements: flowcharts within statecharts, described earlier; sub-statecharts; critical regions, and enumerated flowchart visual switch polygons.

### 2.8.1. Sub-statecharts

Sub-statecharts are a useful tool for describing statechart components within other statecharts and statechart reuse. A sub-statechart is a special state that represents an external statechart. For example, consider the statechart in Figure 2.15a. It references two external statechart objects, one of class TLC1 and the other of class TLC2, where the class name is the name given to the statechart class when the corresponding statecharts generate code. Figure 2.15b shows the statechart for TLC1.

Sub-statecharts enable you to reuse statechart components. For example, in Figure 2.15a, rather than copying the TLC1 statechart and pasting it inside the state, it is simpler to simply reference it using the sub-statechart notation. This is especially useful if TLC1 is designed by a separate team or if it is used in multiple locations or statecharts (or both). Stated differently, a sub-statechart is similar to a coarse state, except that the content of a sub-statechart is described in a separate statechart diagram file and therefore requires separate code generation.

In Figure 2.15a, the statechart, which is denoted as the primary statechart, boots up in the Init state. Afterward, following an *Ev1* event, sub-statechart state *TLC1* is entered. Consequently, an object of type *TLC1* boots up in its initial state (State-1 of Figure 2.15b) and begins operation. This sub-statechart object is now called *live*; it will be live as long as the primary statechart is in the *TLC1* sub-statechart state.

All live sub-statechart objects within a primary statechart receive their event information from the primary. For example, say event *Ev2* fired while the primary statechart in Figure 2.15a is in the sub-statechart state *TLC1*. The primary statechart then invokes the event dispatcher of its *TLC1* sub-statechart object and passes it the current event name, i.e., *Ev2*. Consequently, the sub-statechart object might fire a transition and change states.

As illustrated in Figure 2.15, the StateRover distinguishes between two types of transitions issuing from a sub-statechart state. The first is a transition with an event, such as the transition annotated with the event *Ev1* in Figure 2.15a. If *Ev1* is sensed while the statechart is in state *TLC1*, the statechart will, as expected, leave the *TLC1* state and its nested states and the next state will be state A. The sub-statechart object is then said to be *aborted*. The second type is a null-event transition, such as the transition leading to the *Green* state in Figure 2.15a. This transition fires only after the embedded statechart in Figure 2.15b reaches a *terminal* state.

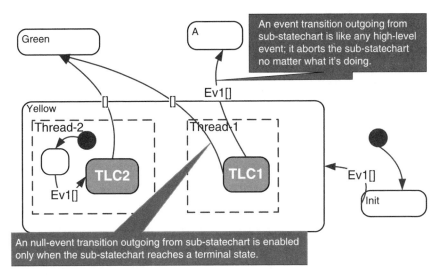

a. A primary statechart with
two sub-statechart states.

b. The TLC1 statechart embedded
as a substatechart in the
primary statechart.

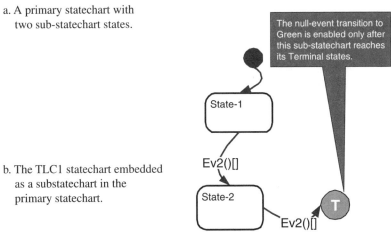

FIGURES  2.15a–b.

The following scenarios illustrate sub-statechart behavior within a primary statechart for the example shown in Figure 2.15:

1.  Events sensed by the primary statechart after boot up are, in order, *Ev1, Ev2, Ev2*. After sensing *Ev1*, the primary statechart transitions from *Init* to *Yellow*, thereby initializing the

sub-statechart *TLC1* to *State-1*. While in state *TLC1*, the primary senses two successive *Ev2* events. The sub-statechart transitions to *State-2* after sensing the first *Ev2* event. Then, after the second *Ev2* event, the sub-statechart transitions to the terminal state. This causes the primary statechart to transition to the *Green* state in the same cycle.

2. Events sensed by the primary statechart after boot up are, in order, *Ev1, Ev2, Ev1*. After sensing the first *Ev1*, the primary statechart transitions from *Init* to *Yellow*, thereby initializing the sub-statechart *TLC1* to *State-1*. As in the previous scenario, the *TLC1* sub-statechart transitions to *State-2* after sensing the *Ev2* event. The next *Ev1* event causes the primary statechart to transition to state *A*. Since the primary is not in state *TLC1*, this transition has the effect of aborting all of *TLC1*'s activities.

Sub-statecharts differ from superstates and coarse states in the following manner:

- The inner content of a superstate is drawn within the scope of the current statechart diagram. As described earlier, the StateRover enables you to draw such refinement either explicitly inside the superstate or on a separate page of the same statechart design, thereby designating the superstate as a coarse state. In contrast, sub-statecharts are references for external statechart diagrams.

- Transitions can cross a superstate or coarse state boundary. The reason is that superstates and coarse states reference states from the current statechart design. In contrast, the only way to "move into" a sub-statechart is to enter the sub-statechart state in the primary statechart, causing it to boot-up in its default state. Also, the only way to "move out of" a sub-statechart is either to have the sub-statechart object transition to a terminal state or to abort the sub-statechart object.

Statechart assertions, described in later chapters, make extensive use of the sub-statechart notation. In fact, a statechart assertion is no more than a sub-statechart object with some additional Boolean information.

## 2.8.2. Enumerated Flowchart Visual Switch Polygons

You can use flowchart visual switch polygons either for Boolean branching or for enumerated type branching, just like a switch statement in Java; hence, the name visual switch. Indeed, in the enumerated type case, the StateRover's code generator generates a switch statement for the visual switch.

Figure 2.16a illustrates a Boolean visual switch, and Figure 2.16b an enumerated visual switch. Note that the transition without an enumerated value serves as the default of the switch statement.

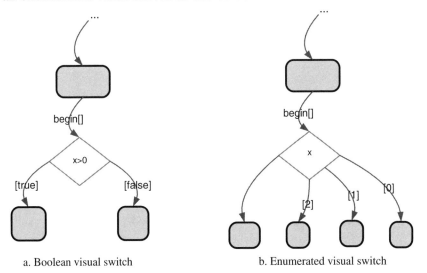

a. Boolean visual switch          b. Enumerated visual switch

FIGURES 2.16a–b   Visual switch.

### 2.8.3. Parameterized Events

Figure 2.17 shows a portion of the TLC statechart with a parameterized event transition. The transition specifies that if event *newCar* is received along with the specified list of arguments (i.e., a *Car* and a *Rolls* object) when the TLC is in state *Count_0*, the transition from state *Count_0* to state *Count_1* is enabled.

Note that this event is different from the *newCar* event, which has no parameters, or the event *newCar* with any argument list that has a signature that differs from (Car, Rolls). For example, if *myTlc* is an instance object of the TLC statechart, calling *myTlc.newCar(car, rolls)* enables the transition to state *Count_1*, while calling *myTlc. newCar()* enables the transition to state *Count_2*.

Events objects are one method used for passing data into a controller statechart from the outside world.

Note that events are passed down to sub-statecharts together with their associated objects.

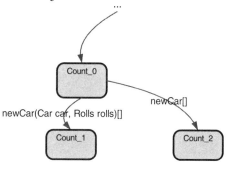

FIGURE 2.17   Transition event with parameters.

### 2.8.4.  Critical Regions

Critical regions describe constraints on simultaneous state visitations within orthogonal states. Consider Figure 2.18. In the absence of a

critical region (Figure 2.18a), when the statechart is in the state configuration <B_l, B_r> and the event *Ev1* is sensed, two transitions fire,
one in each thread, causing <A_l, A_r> to be the next state configuration. The critical region in Figure 2.18b enables at most one thread to
visit its enclosed states. Therefore, either *Thread_left* will transition
to *A_l* or *Thread_right* will transition to *A_r*, but not both. The actual
transition that fires is the one that executes first, based on the fixed
schedule determined by the code generator–i.e., the one whose *if-*
block is the first to execute. We can influence this decision by assigning a higher priority to one transition, using state hierarchy, as shown
in Figure 2.18c, where the *Thread_left* transition has a higher priority
than the competing *Thread_right* transition.

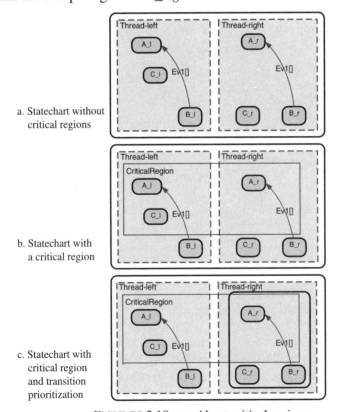

a. Statechart without
   critical regions

b. Statechart with
   a critical region

c. Statechart with
   critical region
   and transition
   prioritization

FIGURES 2.18a–c  About critical regions.

## 2.8.5. Synchronization States

As described earlier, *synchronization states* impose synchronized behavior when concurrent action threads run in an asynchronous mode.

## 2.8.6. Default Events

Default events provide functionality that is similar to that of the *else* part of an *if-then-else* statement in C++ or Java, but it does so for events. Consider Figure 2.19. The default event *eventTRdefault* fires while the statechart is in state *A* if no other event fired a transition; that is, if:

- While the statechart is in state *A*, event *E* occurs but its associated condition guard fails.

- While the statechart is in state *A*, some other event fires (e.g., event *Q*).

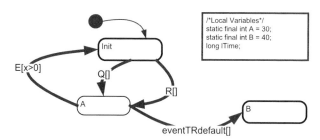

FIGURE 2.19   The default event.

## 2.8.7. Local Variables and Methods

Local variables are member variables wanted as part of the statechart controller class. Local variables are specified using a *local-variables box*. For example, the statechart of Figure 2.11a contains a local-variables box that specified a member variable named *nCnt*. This variable becomes a member variable in the generated class for the statechart controller.

A statechart can have more than one local-variables box and each box can define one or more variables.

Local-variables boxes can be used to define local methods as well as local variables.

## 2.9. Passing Data to a Statechart Controller

The simplest way to pass data objects to a statechart controller is via its constructor. The StateRover lets us specify custom arguments and code to our generated constructor. For example, say we want to pass an *Airplane* object reference to our statechart controller, and then refer to it somewhere inside the statechart. First, define a local variable named *airplane*:

> *Airplane airplane;* // use a local-variables box as described in
> // Section 2.8.7.

Then, pass an Airplane reference named *anAirplane* via the constructor and specify custom code for the constructor:

> *airplane = _anAirplane*;

## 2.10. JUnit Testing of Statechart Objects

JUnit is a popular unit-testing framework for Java that helps create and store tests. It is widely popular among developers and testers. Throughout this book we refer to and use JUnit as a driver for the

creation of test and simulation scenarios. If you are unfamiliar with
JUnit, just think of it as an alternative to a user-created Java file that
stimulates a component under design.

Developers of controller statecharts often use JUnit to unit test
the behavior of their statecharts. Typical JUnit testing is *unit testing*,
where individual Java methods are tested. A statechart, however, is
realized as a collection of methods encapsulated in a class. Therefore,
a convenient test approach, one that amalgamates all event handlers, is
to test the *execTReventDispatcher* event dispatcher method. The following JUnit code illustrates such a JUnit test case for the traffic light
controller in Figure 2.11a; it walks the statechart through a scenario
that visits the sequence of states configurations *Yellow, Green, <Red,
Count_0>, <Red, Count_0>, <Red, Count_0>, Green* (note that the
action takes place in the testExecTReventDispatcher() method):

```
import junit.framework.*;


public class TestTLC extends TestCase {
  private TLC tLC = null;


  protected void setUp() throws Exception {
    super.setUp();
    /**@todo verify the constructors*/
    tLC = new TLC();
  }


  protected void tearDown() throws Exception {
```

```
    tLC = null;

    super.tearDown();

}


                    // This method implements a single
                         test case, i.e., a single scenario.
   public void testExecTReventDispatcher() {
tLC.start();       // sending a start event to the
                    // statechart
tLC.timeout();     // sending a timeout event to the
                    // statechart
tLC.newCar();      // sending a newCar event to the
                    // statechart
tLC.newCar();();   // sending a newCar event to the
                    // statechart
tLC.newCar();();   // sending a newCar event to the
                    // statechart
assertTrue(tLC.isState("Green")); // validate the end
                    // result of this scenario

   }


}
```

Note how JUnit is used not only to feed scenario events and conditions to the statechart but also to verify actions performed by the statechart.

We will use JUnit-based statechart testing in the coming chapters for:

- Simulating statechart assertions.

- Hand-coded testing of statecharts with embedded statechart assertions.

- White box testing of statecharts with embedded statechart assertions.

## 2.11. Statecharts vs. Message Sequence Charts and Scenarios

Clearly, a JUnit test case for a statechart captures a scenario for that statechart. However, a statechart like the traffic light controller in Figure 2.11 embodies a large collection of scenarios. In fact, if the statechart has a loop (such as the loop in the *counter* thread of the traffic light controller), the statechart describes an *infinite* collection of scenarios.

Message sequence charts (MSCs) are a popular UML language used for diagrammatically capturing, communicating, and analyzing *individual* scenarios.

## 2.12. Probabilistic Statecharts

Probabilistic state machines are used in many scientific and engineering applications, such as voice recognition and communications. This section describes a simple method for creating probabilistic statecharts using the code generator described earlier in this chapter.

## 2.12.1. Example: Black Box Environment Modeling Using Probabilistic Statecharts

As an example, let's use the probabilistic statechart in Figure 2.20a to generate events for the traffic light controller in Figure 2.11a. As can be seen, the probabilistic statechart is effectively modeling the controller's environment. This is a form of *black box* testing, where the test suite is created without observing the UML statechart model (the TLC), but rather is created using a model of the statechart's environment.

Figure 2.20b shows the black box testing architecture. The JUnit code below is used as the driver, driving the code generated by the StateRover for the probabilistic statechart, which in turn generates events for the TLC statechart.

The following listing contains Java code for the JUnit driver:

```java
import junit.framework.*;


public class TestTLCgenerator extends TestCase {
private TLCgenerator tLCgenerator = null;
                    // probabilistic statechart test
                    // generator
  private TLC sut; // the statechart under test


  protected void setUp() throws Exception {
    super.setUp();
    tLCgenerator = new TLCgenerator();
    sut = new TLC();
    tLCgenerator.myTlc = sut; /* point the
```

```
    probabilistic generator to send tests to
    the statechart under test*/
  }


  protected void tearDown() throws Exception {
    tLCgenerator = null;
    sut = null;
    super.tearDown();
  }


public void testExecTReventDiapatcher() {
      // a test suite with 100 tests, each test is 50
      // cycles long
      for (int nTest = 0; nTest < 100; nTest++) {
      for (int nCycle = 0; nCycle < 50; nCycle++) {
          double nRand = Math.random();
          tLCgenerator.P100();
          if (nRand < 0.25) tLCgenerator.P25();
          if (nRand < 0.75) tLCgenerator.P75();
      }
      // before running a new test, reset the
      // controller and the test generator
          tLCgenerator.execTRreset();
          sut.execTRreset();
      }
    }
}
```

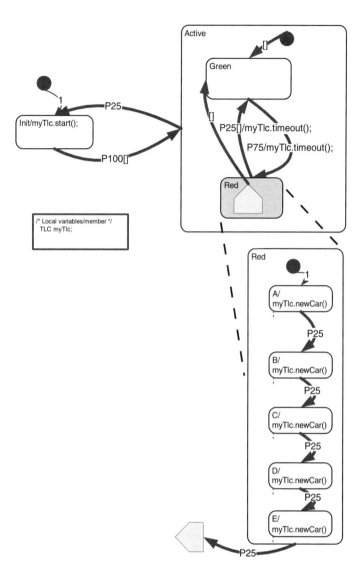

FIGURE 2.20a   A probabilistic statechart.

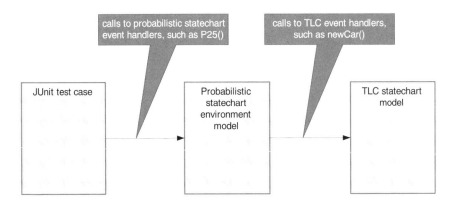

FIGURE 2.20b   Black-box testing architecture.

# Chapter 3

# Academic Specification Languages for Reactive Systems

In this chapter we will consider formal specifications of reactive systems. By formal we mean a specification that is unambiguous and that a computer can understand to the degree that it can execute the specifications.

Historically, formal specification was used to mathematically prove the correctness of a computer program with respect to the **formal** requirement. By providing a mathematical proof, we avoid the need for exhaustive testing and coverage is absolute. However, as we will discuss in Section 5.7.1 formal verification methods have failed to deliver on their promise and in practice work only for small or restricted programs.

In addition, an Achilles heel of classical formal methods has been the specification language and specification process. For example, while it is widely agreed that a natural language cannot be considered a good specification language, in practice it is used for that purpose. This book attempts to address that concern, and Chapter 5 describes a process for developing specifications that uses natural language. In addition, classical formal specification languages deviate all too often in their look and feel from standard modeling languages (UML in particular). That reduces their acceptance in the industry, especially since we can specify the same requirements using UML-like specifications, as we do in this book using statechart assertions.

A note about terminology: in the classical formal methods literature a formal specification is often called a *correctness property*; others, however, often call it a *temporal rule*. Also, when logic is used for specification, a correctness property is often called a *logic formula*. In this book we will refer to formal specifications as *temporal assertions,* or *assertions* for short. Indeed, temporal assertions are used in a manner that resembles the way that Java, C++ or JML assertions are used. Nevertheless, they are different, as we will see in Section 4.5.

## 3.1. Natural Language Specifications

Natural language (NL) is not considered a formal specification language primarily for two reasons. First, in its current state natural language processing (NLP) technology is not capable of machine understanding of nontrivial NL. In other words, although a machine can read and parse NL, and can also synthesize voice for NL, a computer doesn't really know, in the general case, what the NL means. Second, NL is rather ambiguous, as illustrated by the following NL requirement and Figure 3.1, which demonstrates a particular scenario:

**R3.1**: *Whenever* x > 0 *then* y > 0 *should follow within 15 seconds afterward.*

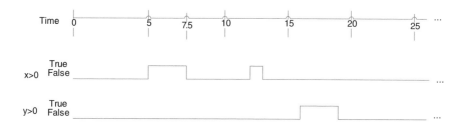

FIGURE 3.1  A scenario for requirement R3.1.

The first ambiguity is rather simple: when do we consider $x > 0$ (or $y > 0$) to occur? Is it when it changes from $x \leq 0$ to $x > 0$ (e.g., time 5 in Figure 3.1), or is it every time when x > 0 is true (e.g., every time stamp between time 5 and time 7.5 in Figure 3.1). Say we resolve this ambiguity by choosing the first option; let's denote the transition of the condition x > 0 (y > 0) from true to false and from false to true as the *value change events*. R3.1 now becomes:

**R3.1**: *Whenever* x > 0 *change event occurs, then the* y > 0 *change event should follow within 15 seconds afterward.*

The second ambiguity is then more subtle: does the R3.1 NL specification consider the scenario of Figure 3.1 as good (acceptable) or bad (unacceptable)? The first interpretation, perhaps the one most programmers believe, is that R3.1a considers the scenario good because $y > 0$ happened within 15 seconds of all instances of the $x > 0$ value change event. A different interpretation, however, is that there must be a unique $y > 0$ change event for every $x > 0$ change event. Given this interpretation, the scenario in Figure 3.1 is bad for R3.1a because it contains only one $y > 0$ change event for both $x > 0$ change events.

Ambiguity is obviously a concern as far as using NL specifications to verify any safety-critical system. Nevertheless, NL is a popular first step in the specification process. It is a primary means by which human beings phrase, evaluate, and communicate specifications. Therefore, inevitably, we need a process for dealing with specifications that originate in NL.

Moreover, the source of ambiguity in the R3.1a NL specification is not necessarily in the language. Say that we used some formal specification language to specify R3.1b, a variant of R3.1a written in an unambiguous manner that is consistent with the first interpretation. But did the developer who wrote R3.1b even consider the second interpretation? He might not have if he did not think through the assertion for the scenario in Figure 3.1.

We will consider this important aspect of requirement construction and simulation in Section 5.3.4.

## 3.2. Using Specification Languages for Runtime Monitoring

In this book, we will use formal specification assertions as an oracle for classifying scenarios as *good* (*acceptable*) or *bad* (*not acceptable*). For example, consider Figure 3.2a. The box represents a specification assertion written in Linear Time Temporal Logic (LTL), a specification language we will discuss in this chapter. The specification assertion (listed inside the box) is concerned with a domain of discourse that consists of the conditions $x > 0$, $y > 0$, $y < 0$, and $y == 0$. An executable version of the assertion is used to evaluate input scenarios and to classify them as good or bad, as illustrated in Figure 3.2a. This is the essence of *runtime execution monitoring* (*REM*), also known as *runtime verification (RV)*.

Rather than mathematically proving the correctness of an assertion for every possible run of the system, as classical formal methods do, REM simply executes the specification and notifies the user or the enclosing test suite manager of the success or failure of the specification vis-à-vis that particular execution. This is the reason for calling specifications *assertions*. In fact, our assertions are functionally similar to Java assertions except that they describe behaviors that are more complex. We will discuss the distinction between Java assertions and temporal assertions in the sequel.

Figure 3.2 illustrates two primary methods that we will consider for writing formal specification assertions for REM. The first uses a formal logic called temporal logic (Figure 3.2a); the other is automata-based (Figure 3.2b). In fact, we will be using statecharts for specification.

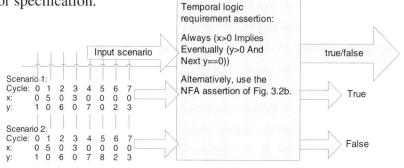

a. Using a formal specification assertion for run-time verification

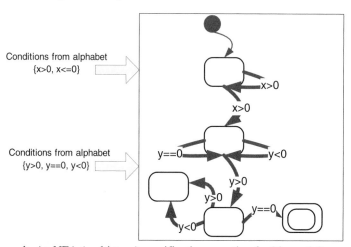

b. An NFA (multi-tape) specification assertion for Figure 3.2a

FIGURES 3.2a–b.

## 3.3. Linear-time Temporal Logic (LTL)

### 3.3.1. Background

Linear-time temporal logic has received considerable attention in academia in the last quarter century. It has been suggested, for reasons that are not always clear, as the preferred language for the specification of "correctness properties" for a system or software component. Classical formal verification methods, such as model checking or theorem proving, use these correctness properties together with a mathematical model of the component to produce either a mathematical proof of correctness or a counterexample.

Classical formal methods suffer from numerous limitations that were discussed in the preface to this book. With a simpler verification method in mind (REM), we can focus on the primary aim of any specification language, which is to specify properties, or assertions, of interest. For that reason, we will also investigate extensions of LTL that include features needed by real-life assertions. In particular we are interested in LTL extensions that describe *real-time* constraints and *data* constraints, also known as *time-series* constraints.

Although the name *temporal-logic* suggests that it is primarily concerned with time, that is not the case. Temporal logic in general, and LTL in particular, are actually primarily about *order* and *sequencing*. LTL assertions describe how basic properties, called propositions, are ordered—i.e., what must be before, or after, what.

### 3.3.2. Informal Syntax and Semantics

Like conventional propositional logic, LTL[1] uses Boolean conditions called *propositions* as basic building blocks. Programmers can

---

[1]   The correct name for the kind of temporal logic discussed in this book is propositional linear-time temporal logic, or PLTL.

read the Boolean condition part of the following Java if-statement:

```
if (x>0 && (!y>0 || z>0)) {...}
```

The Boolean condition x>0 && (!y>0 || z>0) uses three propositions—$x > 0$, $y > 0$, and $z > 0$—as well as the logical *And*, *Or* (&&, ||) binary operators and the unary negation operator. Given particular truth-assignment values for the Boolean propositions, we can evaluate whether the condition is true or false. For example, if $x > 0$ is true, $y > 0$ is false, and $z > 0$ is true, the condition above is true. We then say that the logical formula *evaluated to* true.

Unlike propositional logic conditions, LTL formulae are not evaluated based on a set of truth assignments for Boolean propositions. Rather, an LTL formula is evaluated given a *sequence* of truth assignments for Boolean propositions. In other words, LTL formulae evaluate to true or false given a sequence of truth assignments for propositions, such as the following sequence, denoted $S_1$, of length four:

---

First cycle: x>0 is true, and y>0 is false,
Second cycle: x>0 is true, and y>0 is false,
Third cycle: x>0 is true, and y>0 is false,
Fourth cycle: x>0 is false, and y>0 is true.

---

LTL syntax is similar to propositional logic except that it also contains four *temporal* operators. Hence, the propositional-logic formula described earlier is a special case of an LTL formula that has no temporal operators. Think of it as observing the values of the propositions *in the present*, whereas the temporal operators consider the values of the same propositions in the future or the past, or both. The four temporal operators are described below. A key concept of their description is that of a *cycle*, which we will discuss in Section 3.3.5.

1. *Always (Henceforth) p*, where *p* is another LTL formula, evaluates to true if and only if *p* is true *always*—that is, *every* cycle, now and in the future. Hence, for example, *Always (Not AlarmOn)* is true if *AlarmOn* is never true. Figure 3.3a illustrates the evaluation of *Always p*. The leftmost bar indicates the point in time that evaluation begins. Thereafter, *p* must be true (1) for *Always p* to evaluate to true. It suffices that if *p* is false (0) even once, *Always p* will evaluate to false.

2. *Eventually p*, where *p* is another LTL formula, evaluates to true if and only if *p* is true in *some* cycle now or in the future. Hence, for example, *Eventually AlarmOn* is true if *AlarmOn* is true now or will be true sometime in the future. Note that once *AlarmOn* becomes true, its value afterward is inconsequential to the evaluation of *Eventually AlarmOn*. Figure 3.3b illustrates the evaluation of *Eventually p*. Again, the leftmost bar indicates the point in time that evaluation begins. Thereafter, *if p is* true (1) at least once, *Eventually p* will evaluate to true. If, however, *p* is never true, *Eventually p* will evaluate to false.

3. *p Until q*, where *p* and *q* are other LTL formulae, evaluates to true if and only if *p* is true every cycle until *q* is true sometime, either now or in the future. Figure 3.3c illustrates the evaluation of *p Until q*. The leftmost bar indicates the point in time that evaluation begins. Thereafter, *if p is* repeatedly true (1) until a later time when *q* becomes true, *p Until q* evaluates to true. If *p* becomes false before *q* becomes true,

*p Until q* evaluates to false. There is a third possibility, however, which is that *p* remains true but *q* never becomes true.
This possibility leads to two variants of the *Until* operator:
the *Weak Until* operator, denoted *W*, and the *Strong Until*
operator, denoted *U*. With a *pUq* formula—i.e., using a
strong until–*q* must occur or else the formula evaluates to
false. Hence if *p* is true forever but *q* is never true, *p U q* is
false. On the other hand, a *pWq* formula—i.e., one that uses a
weak until–evaluates to true under the same circumstances.

4. *Next p*, where p is another LTL formula, evaluates to true
   if and only if *p* is true in the next cycle. Figure 3.3d illustrates the evaluation of *Next p*. The interesting case is the last
   cycle, when no next-cycle exists. In such a case, we arbitrarily define *Next p* to succeed. However, this assumption is
   sometimes modified, as in the case of the recurrence equations below.

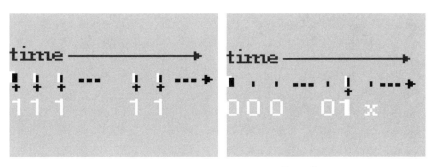

a. Always p ( ).                              b. Eventually p (◊)

FIGURES 3.3a–b.

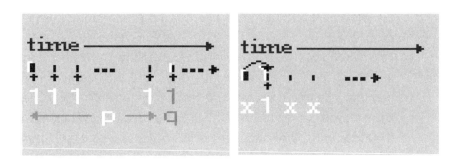

c. p Until q.                              d. Next p(0).

FIGURES 3.3c–d.

LTL is more often than not used in a nested manner, such as *Always ( x > 0 Implies ((Not y > 0) U (Not x > 0)) )*. This formula represents the NL requirement *whenever x > 0 then no y > 0 until x ≤ 0 sometime later*.

### 3.3.3. LTL as a Formal Language Recognizer

LTL can be thought of as a multitape acceptor, similar to multitape FA, where propositions are thought of as alphabet symbols. For example, the LTL formula in Figure 3.2a behaves like a two-tape acceptor: one tape consists of letters from an alphabet, $\Sigma_1$, of two symbols: one represents the condition $x > 0$; the other, the inverse condition. The second tape uses an alphabet, $\Sigma_2$, of three symbols: one represents the condition $y > 0$; the second, the condition $y == 0$; and the third, the condition $y < 0$. Note how symbols in the alphabet for a particular tape always represent mutually *exclusive* conditions.

Hence, an LTL formula can be considered a formal language recognizer, just like a multitape FA, discussed in Chapter 1. The LTL formula in Figure 3.2 recognizes a language, L, over the Cartesian product $\Sigma_1 \times \Sigma_2$. Any string s from $\Sigma_1 \times \Sigma_2$ is in L if and only if the LTL formula evaluates to true for the input string *s*. The sequence $S_1$ described earlier is an example of a string in L.

This interpretation of LTL specifications as language recognizers lets us substitute LTL assertions with FA assertions and statechart assertions, as recommended in the next chapter.

### 3.3.4. LTL Tautologies and Recurrence Equations

Some interesting tautologies exist for LTL, such as:

- *Always p == Not Eventually Not p*, because clearly *p* is always true if and only if *Not p* never happens.

- *Eventually p == Not Always Not p.*

- Always p == p W false.

- *Eventually p == true U p.*

- *pUq = (pWq) And (Eventually q).*

- *pWq = (pUq) Or (Always p).*

The following *recurrence equations* are often used by LTL code generators and interpreters:

- *Always p = p And Next Always p.*

- *Eventually p = p Or Next Eventually p.* In this case we define the Next operator to fail on the last cycle.

- *pWq = q Or (p And Next pWq).*

- *pUq = q Or (p And Next pUq).* In this case, too, we define the Next operator to fail on the last cycle.

These recurrence equations are important for two primary reasons. First, they are used by execution engines, such as the TemporalRover discussed later in this chapter. Second, they reveal some fundamental aspects of the language, as follows. Consider the LTL formula *Always p*, where *p* is a basic proposition, such as *x > 0*. Using the first recurrence we have the following *unwinding*:

*Always p = p And Next Always p = p and Next (p And Always Next p) …*

*= p And Next p And Next Next p And Next Next Next p And Next Next Next Next p … …*

In other words, LTL operators are really based on repeated *Next*'s; that is, any LTL formula, once unwinded, becomes an infinite propositional logic formula augmented with a single temporal operator, *Next*.

As we will see in Section 3.4, Interval Temporal Logic (ITL), which is a different type of temporal logic, uses a concatenation operator as its basic temporal operator.

### 3.3.5. Executing LTL Assertions and the Notion of a Cycle

Let's evaluate the LTL formula *Always ( x > 0 Implies ((Not y > 0) W (Not x > 0)) )* given the input sequence $S_1$. First, we convert the formula into a more manageable equivalent, using the well known tautology *a Implies b == ((Not a) Or b)*:

*Always ( (Not x>0) Or ((Not y>0) W (Not x>0)) )*

LTL evaluation proceeds from the outside of the formula inward. Hence, we see the LTL formula *Always p*, where *p* is *(Not x > 0)*

*Or ((Not y > 0) W (Not x > 0))*. For *Always p* to be true, *p* must be true in all cycles. Obviously, *p* is true in the fourth cycle, since x > 0 is false then. However, the evaluation of *p* in the first, second, and third cycles is more involved. In those three cycles, *(Not y > 0) W (Not x > 0)* evaluates to true because *Not y > 0* holds true (*y > 0* being false) until the fourth cycle in which *Not x > 0* is true (*x > 0* being false).

Hence, since *p* is true in all cycles being considered, the LTL formula *Always p* evaluates to true for the input sequence $S_1$.

With REM, which we have already discussed briefly, a formal specification is executed in tandem with a program or model. The assertion is first associated with a certain line of code of the program, as discussed in Section 3.3.12. This line of code defines the notion of a cycle in the LTL assertion, thus: every time this line of code is reached is defined as a new cycle and the LTL assertion is re-evaluated for that cycle (see Section 3.3.12 for more information). Hence, REM of an assertion is no more than repeated evaluation of the assertion, one cycle at a time. Each cycle, basic propositions, such as the conditions x > 0 and y > 0 of Figure 3.2, are sampled from the program space and sent to the LTL formula, which is then re-evaluated.

Consider for example the process of performing REM of the assertion:

*Always ( x>0 Implies ((Not y>0) W (Not x>0)) )*

Say the executing program is monitored to exhibit the behavior of sequence $S_1$. After the first cycle is monitored (*x > 0* is true and *y > 0* is false), the LTL formula is evaluated as if the test sequence consists of only one cycle. It then publishes its decision (assertion succeeded), but with a grain of salt, because this decision is not necessarily final; if the test is extended the decision might change. After the second cycle (again, $x > 0$ is true and $y > 0$ is false), the formula is re-evaluated assuming a test-sequence of two cycles, and the decision is published again. This process continues every cycle for the duration of the program.

As we said, the cycle by cycle decisions of the monitor are often transient. For example, the formula *Eventually AlarmOn* publishes a decision *false* as long as the sampled basic proposition *AlarmOn* is false. However, once *AlarmOn* becomes true, the assertion *Eventually AlarmOn* succeeds (evaluated to true). Also, interestingly, this last evaluation is no longer transient, because no future change to *AlarmOn* changes the fact that *AlarmOn* was true at least once.

Likewise, the evaluation result for the assertion *Always (Not AlarmOn)* is transient (as true) as long as *Not AlarmOn* is true and until *Not AlarmOn* becomes false, if ever. If it does, the assertion fails and retains this status no matter how *AlarmOn* changes in the future.

The TemporalRover and DBRover LTL tools, discussed below, actually provide two bits of information as an LTL evaluation result. The first is the logical success/fail (good/bad) result every cycle. The second indicates whether or not the result is transient.

### 3.3.6. LTL with Real-Time Constraints

Consider the LTL assertion *Always (p Implies Eventually q)*, which requires that whenever condition *p* is true, condition *q* should be true then or sometime later. This kind of an assertion is rather useless in practice because it places no constraint on the eventual possibility. If *p* is "911-emergency-call-ringing" and *q* is "911-operator-answers-call," the assertion will be satisfied even if the operator answers the call a year after the emergency occurred. Clearly, in real life almost every promise for the future must have an upper-bound constraint.

The need for real-time constraints occurs elsewhere in LTL as well. For example, a requirement that a traffic light must be Red for at least 1 minute is considered a lower-bound constraint on the Always operator.

Metric Temporal Logic (MTL) is suited for real-time constraints. With MTL, each of the four temporal LTL operators is optionally augmented with a real-time constraint, which can be an upper bound, lower bound, or both.

Consider the MTL assertion $Eventually_{\leq 10}$ $alarmOff$. It specifies that $alarmOff$ should be true in up to 10 real-time units.

Like LTL, MTL can be nested, as in the assertion:

*Always account<0 Implies*

*(Eventually $_{\leq 1\text{-}week}$ Always $_{\leq 1\text{-}month}$ account>0)*

The assertion specifies that whenever a bank account is in the red it should recover within one week, and recovery is specified as the account being in the black continuously for an entire month. Note that recovery needs to begin within a week, but the end of the recovery period will clearly be after an entire month, possibly at the end of a month and one week. This requirement can fail in two main ways. First, a recovery might not begin within a week. Second, a recovery might be incomplete, such as the account being in the black for only two weeks rather than a month as specified.

## 3.3.7. LTL with Time Series Constraints

It is difficult if not impossible to represent three types of properties in LTL: stability, monotonicity, and min/max over time. LTL with time-series constraints (TLS) is an extension of LTL created for that purpose. The following TLS *cruise-control* stability requirement, requiring speed to be 95% stable while cruise is set and not changed, illustrates the use of TLS:

```
Always (

   (CruiseSet Implies  speed*0.95<speed' And
   speed'<speed*1.05)

   Until $speed$

   (cruiseChange || cruiseOff)

)
```

In this assertion, *speed* is a temporal data variable associated with the *Until* temporal operator, as indicated by the syntax *Until $speed$*. This association implies that every time the *Until* operator begins its evaluation, the speed value is sampled and preserved in the *speed* variable of this instance of the *Until* operator; this value is referred to as the pivot value for this *Until* operator instance. Future speed values used by this particular evaluation of the *Until* statement are referred to using the prime notation (*speed'*) and are called primed values. Hence, if speed is 100 Kmh when *cruiseSet* is true, the pivot value for speed is 100, and every subsequent speed value is referred to as *speed'* and must be within 95% and 105% of the (pivot) speed.

### 3.3.8. Past Time LTL

Past-time Temporal Logic (PTL) extends LTL with four past time operators, which correspond to the LTL operators:

1. *Always-in-the-past p*, where *p* is another PTL formula. It evaluates to true if and only if *p* is true in the present cycle and in every preceding cycle since the beginning of the input sequence.

2. *Sometime-in-the-past p*, where *p* is another PTL formula. It evaluates to true if and only if *p* is true in the present cycle or in an earlier cycle.

3. *p Since q*, where *p* and *q* are other PTL formulae. It evaluates to true if and only if *p* is true every cycle since sometime in the past when *q* was true. As with the *Until* operator, two variants of the *since* operator exist: the *Strong Since*, which requires *q* to exist, and the *Weak Since*, which does not require *q* to be true any time in the past.

4. *Previous p*, where *p* is another PTL formula. It evaluates to true if and only if *p* was true in the previous cycle. As with the *Next* operator, we arbitrarily say that *Previous* p succeeds on the very first cycle, although there is no previous cycle.

Every PTL formula can be converted into LTL. For example:

*Always (p Implies (q Since r)) ==*

*Always ((Eventually p) Implies ((Not p) U (r And (q U p)) ))*

This equivalence is based on the following informal observation:

*Always (p Implies (q Since r)) == Always ((Eventually p) Implies (r exists before this p and q is true from that point until this p occurs)*

This example illustrates that, although past temporal operators are not necessary from the standpoint of descriptive power, they contribute to the readability of the temporal specification.

### 3.3.9. Infinite-Sequence Semantics of LTL and Automata

In Chapter 5 we will briefly describe model checking (MC). MC and other heavyweight formal methods use mathematical reasoning to prove that a model (a statechart in our case) conforms to a formal specification $p$: namely, that if MC says that $p$ is conformed to, no more testing is needed for that assertion. Classical formal methods typically use LTL with an *infinite-sequence* semantics, which defines the meaning of the four future-time operators given an infinite number of cycles. Infinite-sequence semantics is similar to finite-sequence semantics, described earlier, except that the *Next* operator is always well defined because there is always a next cycle.

Clearly, no test or execution of the system runs for an infinite number of cycles. That is the reason that in this book we have little use for such semantics. However, when proving a mathematical theorem without actually executing the program, we cannot make any assumption about the length of an execution of the program. Therefore, the most convenient mathematical representation is simply to assume that the execution is of infinite length.

LTL is not alone in needing readjustment for infinite sequences; automata and statecharts too need to be defined over infinite input sequences. A requirement is no longer defined as a set of finite strings. Rather, it is a set of infinitely long scenarios (strings) of letters from the alphabet. So instead of $\Sigma^*$, we now have $\Sigma^\omega$, the set of all infinitely long scenarios that can be constructed from letters of

Σ. Automata definitions are also somewhat different now. For example, DFA over infinite scenarios, now called ω-DFA, has a different acceptance criterion, tailored for infinite scenarios, such as: if an infinite computation visits some final states an infinite number of times, the (infinite) input sequence is accepted.

## 3.3.10.  LTL Expressive Power and Succinctness

LTL is less expressive than FA; that is, it cannot express all regular languages. The following NL requirement, for example, cannot be expressed in LTL:

**R3.3.10**: *x > 0 must be true in every even cycle.*

It is tempting to think of the following as an equivalent LTL specification:

*x>0 And Always ( (x>0 Implies Next (Not x>0))  And ((Not x>0) Implies Next x>0)  )*

However, it is in fact overkill, because it evaluates to false on the following sequence of *x*'s: 1,0,1,1, whereas it satisfies the NL requirement (we denote the first cycle as cycle 0).

Similarly,

*x>0 And Always (x>0 Implies Next Next x>0)*

evaluates to false on the following sequence of *x*'s: 1 1 1 0 whereas it satisfies the NL requirement.

In contrast, the DFA of Figure 3.4 implements R3.3.10 correctly.

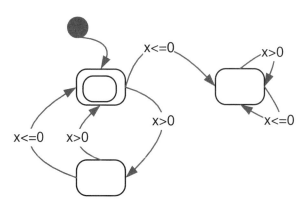

FIGURE 3.4.

LTL's expressive power is known to be the same as a pure sub-set of regular requirements known as *star-free regular requirements* that can be described as follows. Say you were only allowed to use a restricted kind of FA (DFA or NFA) for specification, called *loop-free FA*, in which the underlying graph never contained loops of length greater than one; for example, the DFA of Figure 3.4 has a loop of length two and would therefore be illegal. In other words, the only allowable loops are self-loops from a state right back to itself. The set of requirements specifiable by loop-free FA is the star-free regular requirement set.

One way to think about how LTL compares with FA follows from the LTL to AFA conversion described in the next section, which presents a specific pattern of AFA that corresponds to LTL. In other words, LTL corresponds to, or has the same descriptive power as, a subset of all AFA that use the pattern shown there.

With respect to succinctness, LTL is similar to AFA vis-à-vis the lower bounds discussed in chapter 1 (Figure 1.6+0.5). In other words, for those requirements that are describable in LTL, some are exponentially more succinct than the equivalent NFA, for example. The intuitive reason behind this relationship is that LTL contains the three primary Boolean operators (*And*, *Or*, and *Not*) while NFA are only capable of performing the equivalent of set-union, which is like a logical Or. From logic theory we know that any logical property expressible with the three operators is also expressible using only two. For example, consider the *property (p And Not (q Or r))*; using *And* and *Or* alone, it can be written as *(p And (!q And !r))*, where *!q* and *!r* are considered basic propositions. AFA have the ability to perform two necessary operations, while NFA on the other hand are weaker, and therefore require many more states to represent the same requirement.

The bottom line is that some properties expressible using FA are not expressible using LTL. Since LTL is so closely related to AFA, for those properties expressible in LTL we expect some LTL properties to be exponentially more succinct than the equivalent NFA and double exponentially more succinct than DFA.

Rather than using AFA to diagrammatically represent temporal assertions, it is preferable to use nondeterministic statecharts, with orthogonality replacing AFA *and* nodes. Section 3.5 describes the application of deterministic and nondeterministic statecharts for specification.

### 3.3.11. From LTL to FA

The first step of the conversion is to push all negations into the formula. This is done using DeMorgan's laws for Boolean operators, such as *Not (p Or q) == (Not p) And (Not q)*, and LTL equations for temporal operators, such as *Always p == Not Eventually Not p*.

The next step is illustrated in Figure 3.5, which shows the pattern used to recursively convert LTL into ε-AFA based on the recurrence equations listed earlier.

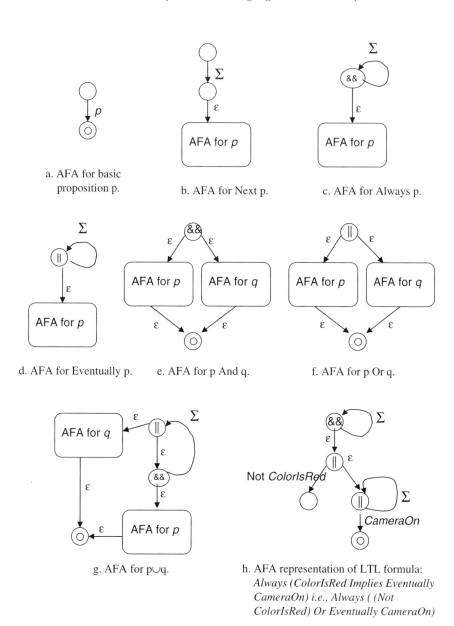

FIGURES 3.5a–h   LTL to ε-AFA conversion.

## 3.3.12.  LTL Tools

The Temporal Rover is an LTL code generator. It supports all the extensions described earlier. As illustrated in Figure 3.6, the Temporal Rover parses a C, C++, or Java source code file that includes LTL assertions written inside specially tagged comments. These comments contain one or more temporal assertions. For example, the temporal assertion in the figure asserts that whenever the subject traffic-light controller is in the Green light mode, the *Camera* is off. The Temporal Rover considers code written inside braces a basic proposition—that is, a Boolean condition. For example, in the figure, a basic proposition is the Boolean condition Camera==0. This condition is executed verbatim; it returns a true or false value, which is then used for the evaluation of the whole temporal assertion.

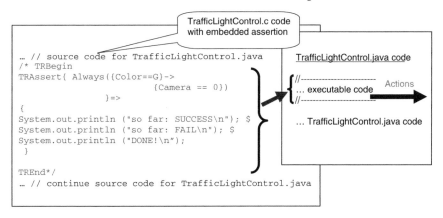

FIGURES 3.6   The Temporal Rover code generation process.

As a result of the Temporal Rover's code generation process, source code replaces temporal assertions within comments, as shown. The new source code file is then used in practice. While it is executing, whenever the assertion code is reached, the temporal assertion is re-evaluated given the current values of the basic propositions.

Each time the assertion code is reached is considered a new cycle for the purpose of temporal evaluation. For example, in Figure 3.6, say that when the assertion code is reached the first time Color==G and Camera==0 are true, hence the temporal assertion evaluates to true that cycle. When the assertion code is reached next time around it is considered the second cycle. Now say that this time, Color==G is true and Camera==0 is false: therefore the temporal assertion fails.

A temporal assertion written for the Temporal Rover also contains custom actions. Figure 3.6 shows three custom actions: one prints a message every cycle in which the assertion succeeds, the second prints a message every cycle in which the assertion fails, and the third prints a message if the assertion result is no longer transient.

The DBRover is an LTL tool based on the Temporal Rover code generator. It enables us to graphically simulate LTL assertions, as well as providing a GUI for specifying scenarios and presenting the evaluation results for the LTL rules under those scenarios. (We will discuss the motivation for simulating assertions in detail in Chapter 4.3.)

Figure 3.7 shows the DBRover simulation of a particular scenario for the assertion:

*Always (Request implies Eventually$_{\leq 1\,minute}$ Ack Until Not Request)*

As shown, we enter a scenario by setting basic propositions to true (1) or false (0) along a timeline. This yields a representation similar to a timing diagram. The DBRover scenario also contains real-time constraints and time-series constraints, which are also shown.

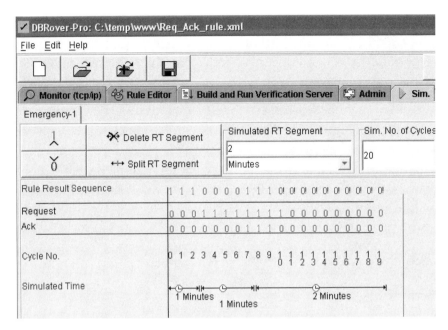

FIGURE 3.7    DBRover simulation of an MTL assertion.

The StateRover statechart tool enables us to associate MTL assertions with statechart states, as illustrated in Figure 3.8, where an MTL assertion is associated with the state *WaitForKeyPressed*. It is therefore quite intuitive to expect that the assertion starts executing every time that state is reached. The notion of a temporal logic cycle is then made to coincide with any event sensed by the statechart, whether or not it changes states. Hence, in the figure *Next* means the arrival time of any event after the entry of the *Wait-ForKeyPressed* state. Note that, by that time, the statechart might be visiting another state.

The StateRover also allows for another interpretation in which the assertion is evaluated only as long as the statechart is visiting *WaitForKeyPressed*.

The StateRover uses the Temporal Rover to generate code for embedded MTL assertions. This code is integrated with the code generated for the statechart, thereby creating a hybrid statechart with an application.

Often an embedded assertion asserts about statechart events. For example, in Figure 3.8 the assertion asserts about the *keyPressed* and *end* events. However, LTL and MTL propositions are really conditions. The StateRover performs automatic namespace translations so that statechart events can be used as basic propositions within assertions.

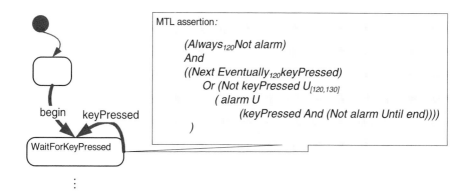

FIGURE 3.8    Temporal logic assertion associated with a UML state.

### 3.3.13.  Pros and Cons of Using LTL Variants for REM

LTL was amongst the first languages used for REM, primarily because it was the specification language of choice for formal methods. In this section we'll consider some of the pros and cons of LTL when applied to REM.

**Pros:**

LTL operators resemble natural language: Always, Eventually, Until, Since, Next, are all taken from natural language. This suggests that it should be quite natural for a person to express natural language specifications in LTL. He could then simulate those properties, correct them if needed, and deploy them for monitoring.

As an example, consider the NL specification *Whenever light is red, light should turn green within 2 minutes*. Four simple **steps** are used to convert it into MTL:

a. Write the NL statement: *always when the light is red, it should turn green within 2 minutes*. In other words, the word *whenever* is substituted with *always*.

b. Rewrite the NL statement as: *always (if the light is red, it should turn green within 2 minutes)*. The purpose of the parenthesis is to remove ambiguities with respect to the scope of the *always*.

c. Rewrite the NL statement as: *always (if the light is red implies that eventually it should turn green within 2 minutes)*. This format inserts an explicit eventually requirement, one which is implicit in the NL statement. This brings the statement one step closer to LTL.

d. Choose *light_is_red* and *light_is_green* as basic propositions, resulting in the MTL assertion *Always (light_is_red Implies Eventually$_{<2min}$ light_is_green)*.

**Cons:**

*Poor readability*: Despite their resemblance to NL requirements, more complex LTL assertions, which are inevitably nested, are hard to read. Since readability is the primary attribute of any computer language, this is all but fatal.

For example, consider the following NL specification for an infusion pump: A session is the interval between *begin* and *end* events. For every such session, *keyPressed* must be repeatedly sensed within two-minute intervals or else an *alarm* must sound within 10 seconds and until *keyPressed* is sensed. Also according to this specification, once *keyPressed* clears the alarm, it should not fire again until the end of the session.

The following MTL assertion captures the requirement using *begin*, *keyPressed*, and *alarm* as basic propositions:

```
Always ( begin Implies

      ( ((begin Or keyPressed) Implies

         ( (Always≤120 Not alarm) And

            ((Next Eventually≤120 keyPressed)

               Or (Not keyPressed U [120,130]

                  (alarm U

                        (keyPressed And (Not alarm

                        Until End))))

            )

         )

      ) U end

   ))
```

Clearly, this assertion is complex and difficult to read. And because it is difficult to read, it will be close to impossible to maintain; in fact, it might not even capture the original intent properly.

The biggest issue with this assertion is its horrific nesting. An LTL proponent could therefore say that LTL is quite handy when nesting is limited to, say, one level, as in *Always (p Implies Eventually q)*. However, given that LTL has only four temporal operators, limiting the level of nesting places a limit on the number of possible formulas, probably a few dozen, that we can write. In fact, we could probably write libraries of clear, well-thought-out NL specifications for all those possible specifications. Moreover, for each such specification we could also write an efficient Java implementation. Certainly that would be a nontrivial project, but it would be a one-time effort that, once complete, will remove the need for LTL—or for any limited specification language for that matter.

Hence, using a specification language only for the "simple specs" actually runs contrary to the need for such a language at all. Stated differently, a specification language is judged by its performance when assertions are complex (and real-life).

Another criticism an LTL advocate could make with respect to the infusion pump example is that, there, nesting is a result of the MTL assertion, mimicking behavior usually associated with a statechart. In fact, why not have a statechart that detects the *begin* event and subsequent *keyPresses*, where *WaitForKeyPressed* is the state in which an assertion should be placed, as illustrated in Figure 3.7? This criticism brings us to the next LTL drawback.

*Unnatural relationship with UML modeling:*

The assertion in Figure 3.8 is associated with the state *WaitForKeyPressed* in a UML statechart, with the expectation that every time the state is entered the assertion will be checked. However, as discussed earlier, after the statechart leaves the *WaitForKeyPressed* state, the assertion is still evaluating, perhaps waiting for a promised eventuality to occur. In other words, although the assertion is written inside a state, it really influences only the initiation of the assertion monitoring, and afterward the assertion is not necessarily associated with any state.

*Events vs. Conditions:*

With LTL, propositions (again, essentially Boolean conditions), are the only basic element. When LTL assertions assert about statechart models those models incorporate both events and Boolean condition guards. For example, the assertion in Figure 3.7 asserts about a proposition named *keyPressed,* which is actually a statechart event. If we want to assert about a property that involves *keyPressed,* someone—person or tool—must convert the event into a Boolean condition, as the StateRover tool does.

The StateRover tailors LTL REM to an event-based system using two adjustments. First, every event *eP* sensed by the REM tool is associated with a basic proposition P. For a given cycle, P evaluates to true if and only if *eP* occurs in that cycle. Next, a constraint is added: in every cycle at most one such proposition may be true. This constraint reflects the simultaneity constraint on events we discussed in Chapter 2.

Having such an event-to-proposition mapping available is convenient but nevertheless, the real issue with LTL still stands: the specification domain (LTL propositions) and the real-life domain (events modeled by statechart transitions) are not the same.

*Weak expressive power:*

LTL is less powerful than FA or plain statecharts. Moreover, as described in Chapter 2, our statechart notation is Java based; i.e., we can write any Java statement as a statechart action, any

Java condition as a statechart transition guard, and any Java method name as a transition event. This makes our statechart notation Turing equivalent.

*Others:*

- LTL is nonvisual, unlike UML.

- LTL is a rather unfamiliar language.

- Even simple artifacts, such as counting, require language extensions.

- LTL is overkill for many simple requirements, in the sense that users don't feel they need the capabilities of a high-end academic specification language to formally write the equivalent of the requirement whenever Request is sensed then within 1 minute Acknowledgement received. Instead, they feel that they could code it quite easily, or if they prefer the visual approach, that they could easily model it as a (deterministic) statechart. As we will see in the next section, many assertions can be expressed using simple, deterministic statecharts, which, using a code generator like the StateRover, result in readable and efficient code. Because LTL is nonvisual and unfamiliar, going through LTL to achieve the same end result seems to many the wrong approach.

- LTL is not necessarily object oriented, unless a code generator generates code that is purely object oriented. Therefore LTL is not necessarily truly encapsulated within the code that uses it.

- LTL is difficult to debug. Even with the advent of a graphical simulator, such as the DBRover, if the assertion does not unambiguously specify the correct behavior, it is difficult to identify which part of the formula is at fault. The only known debugging technique is to break the formula down to smaller, manageable parts; simulate each one alone; and try to deduce from the behavior of the parts where things went wrong with the whole.

- No 1-level recovery technique exists for LTL assertions. Recovery is discussed in Chapter 5.

- There is no tool that performs monitoring of liveness requirements in the strong sense discussed in Section 4.1.6.

Many of these drawbacks apply to other forms of temporal logic, such as branching time temporal logic and interval temporal logic described below. The one negative common denominator all reactive system formal-specification languages other than statechart assertions have is that they differ from the modeling language, i.e., statecharts. This issue has been difficult for nonacademic users to accept.

## 3.4. Other Formal Specification Languages for Reactive Systems

### 3.4.1. Regular Expressions

Regular Expressions (RE, or regex) are textual specifications taken from formal language theory. An RE represents a requirement as a set of scenarios using set notation. RE are constructed recursively from other RE. The basic RE are the empty set; the empty scenario; and a scenario with a single letter, i.e., $\emptyset$, $\varepsilon$, and $a$ for any $a$ in $\Sigma$, respectively. The operations for constructing new RE from smaller ones are:

1. Set union, represented by the + operator. For example *ab* + *aa* represents the set of scenarios {*ab, aa*}.

2. Concatenation, for creating longer sequences from shorter ones. The concatenation of two RE R, and S, denoted as R.S, is the RE that consists of all strings that can be constructed by taking the first part from R and the second part from S. For example if R is *ab* + *aa*, and S is *abbb* + *a*, R.S is *ababbb* + *aba* + *aaabbb* + *aaa*.

3. The *Kleene* star operator*. This operator is like a loop in an FA. The RE R*, where R is an RE, is the infinite union $\varepsilon$ + R + R.R + R.R.R + R.R.R.R +... In other words, R* is the set that consists of all scenarios we can create by concatenating one string in R with others from R, as many times as we want. For example (*ab* + *aa*)* contains *ab*, *aa*, *abaa*, *abab*, *aaab*, *aaaaaaaaa*, *aaabaaab*, etc. Note that unless R is empty, R* is infinite; i.e., it contains infinitely many scenarios (all finite, though).

LTL operators resemble RE operators in the following way. First, we need to think of LTL as a formal language recognizer, as discussed in Section 3.3.3. Hence, an LTL proposition, *p*, now corresponds to a letter from some alphabet, and *Not p* corresponds to a different letter. Now:

- *Always p* is like *p**.

- *Eventually p* is like (*Not p*)**p*(*Not p* + *p*)*, stating that *p* must occur now or sometime in the future and that we don't care what happens afterwards.

- *Next p* is like (*Not p* + *p*)**p*(*Not p* + *p*)*, stating that *p* must occur in the next cycle and we don't care what happens before and after that.

- *p U Q* is like *p\*Q(Not p + p)\**, stating that *p* must be true until *Q* is true (either immediately or in the future) and that we don't care what happens afterwards.

RE are a popular pedagogical tool, but they are limited primarily in their inability to specify set complementation ($\Sigma$\* – R) and intersection (R $\cap$ S) operations. Extended Regular Expressions (EREs) are REs with those two operators added.

EREs suffer from many of the limitations of LTL, as described in Section 3.3.13. For example, EREs are nonvisual, they differ from the UML modeling language they are expected to assert about, and they operate on condition-like letters rather than events.

### 3.4.2. Interval Temporal Logic

Interval Temporal Logic (ITL) can be loosely characterized as a hybrid of regular expressions and logic. Rather than having the *Next* operator as the basic ingredient, as described in Section 3.3.4, ITL uses the *chop* operator (";"), which is but a concatenation, making ITL look a lot like like ERE. The primary distinction between ITL and ERE is that ITL formulas are written in a top-down manner, making ITL somewhat easier to use. This is the reason ITL calls its ";" operator a *chop* operator, which implies top-down breakup of intervals, rather than a *concatenation* operator, which implies a bottom-up construction of sequences.

### 3.4.3. Graphical Interval Logic

Graphical Interval Logic (GIL) is a visual language for temporal specifications that provides an intuitive notation for representing the relative ordering of events in a system using superpositioning of timelines.

Like other languages mentioned in this book, a GIL requirement is a set of legal scenarios. A scenario is described as an interval that is defined by its left and right sides. An interval is represented graphically as a horizontal timeline, with time flowing left to right. The left and right sides of an interval timeline are either basic propositions as in LTL or are defined by other timelines drawn above it. Hence, complex requirements are constructed by super-positioning of simpler ones.

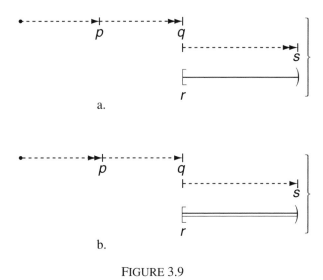

FIGURE 3.9

Consider, for example, Figure 3.9a. The top-most timeline requires that either $p$ not hold (a single arrowhead indicates that $p$ is optional), or that if $p$ holds, then $q$ must hold (a double arrowhead indicates that $q$ is must follow). Note that $p$ is a basic proposition or else it would require a separate timeline. The second timeline states that if $q$ is required to hold, then so does $s$. The third timeline starts and ends where the second timeline started and ended and requires $r$ to hold at the beginning. The third timeline is depicted with a single horizontal line, indicating that it can be empty—namely, that $s$ can hold at the very beginning of the timeline. A double-lined timeline as in Figure 3.9b indicates that the timeline cannot be empty.

Being visual, GIL is an improvement over LTL. Nevertheless, GIL suffers from many of the drawbacks of LTL, such as:

- GIL is not directly related to UML statecharts. In other words, how do we associate GIL assertions with the statecharts they're supposed to monitor?

- GIL is purely propositional (has no events).

- GIL's expressive power is weaker than Java statecharts.

- GIL is unfamiliar, and GIL specifications are difficult to debug, animate and simulate.

- GIL is overkill for simple requirements. In other words, what justifies using a special, unknown specification language rather than using the same language used for modeling?

- GIL has no 1-level recovery mechanism.

### 3.4.4. Computation Tree Logic

Computation Tree Logic (CTL) contains, in addition to the Always and Eventually temporal operators of LTL, two path, or computation, operators: *there-exists-a-computation (E)*, and *all-computations (A)*. The following example illustrates how CTL formulae assert about the allowable computations of a system: **A** *Always x > 0 Implies* **E** *Eventually y > 0* represent the NL requirement: For all system computations C, and for every state S in such a computation in which $x > 0$, there exists a continuation C' of computation C beyond S such that eventually in C' $y > 0$ holds.

With CTL, path and temporal operators must always be used in pairs. For example, *A E x > 0* is illegal in CTL and so is Always

Eventually $x > 0$. CTL* is a temporal logic where the above-mentioned restriction is removed.

Though LTL has no path operators, it is in fact often used with an implicit *A* operator, as in the case of REM. For example, consider the LTL assertion:

*Always (P Implies Eventually Q).*

REM of this assertion checks that every computation satisfies the assertion. This interpretation is the same as writing the assertion in CTL* as:

*A Always (P Implies Eventually Q).*

### 3.4.5. TLCharts

TLCharts are a hybrid of statecharts and MTL. While Figure 3.8 illustrates a statechart with MTL actions, TLCharts also allow MTL conditions. For example, in Figure 3.10, consider the transition *Closed → Init*. This transition's statechart guard *[end{end U begin}]* contains two parts: the Boolean condition *end*, which is the kind of a Boolean condition we expect to see in a statechart transition guard, and the LTL formula *end U begin*. The transition *Closed → Init* is traversed only if both parts evaluate to true. To evaluate the temporal logic part, however, one needs to look into the future to determine that indeed *end* was successively true until a later time when *begin* turned true. While future values of *end* and *begin* are not available in runtime, they are available in test-time if the system execution is recorded. For this reason TLCharts are not used for modeling and code generation, but only for specification and verification. A semiformal syntax and semantics of TLCharts is provided in the Appendix.

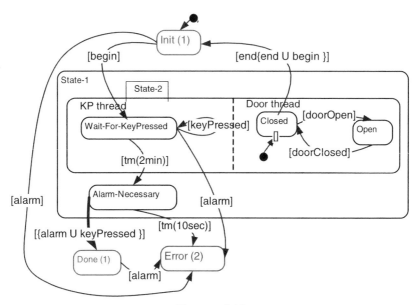

FIGURE 3.10

TLCharts are a superset of both statecharts and MTL. A TLChart with no LTL or MTL guard is a statechart. A TLChart with one state is effectively an LTL or MTL assertion.

# Chapter 4

## Using Statechart Assertions for Formal Specification

## 4.1. Statechart Specification Assertions

While statecharts are part of the UML standard, using statecharts, and particularly nondeterministic statecharts, as assertions is a new concept. Most of the discussion in this chapter uses the StateRover approach and terminology for assertion statecharts. Nevertheless, almost everything we discuss here can be done using other tools, albeit with more manual work.

### 4.1.1. Assertion Statecharts: the Basics

In a nutshell, *statechart specification assertions*—statechart assertions for short—look and feel like the statecharts we've considered in Chapter 2, with two differences:

1. They have a built-in mechanism for indicating Boolean success or failure (true/false), which makes them suitable for formal specification, like LTL.

2. They can be nondeterministic, if desired.

The obvious questions that arise are: *If I already have a stat-echart model, why would I want to add such Boolean information? Who will be using it?* The answer is that the model statechart is not the assertion statechart we will consider here. Rather, our assertion statecharts are separate from the model, although used by it. We will discuss this issue in detail in Chapter 5.

Another obvious question is: *If I already have a model stat-echart, why do I need separate assertion statecharts that say wheth-er the model is doing ok? My model is already doing the job.* That is a good question with multiple answers, all of which we will also discuss in Chapter 5.

Figure 4.1a contains a variation of the TLC in Figure 2.7a, with an embedded assertion called *Assertion1,* shown in Figure 4.1b, which asserts the NL requirement *at most N newCar events should ever be counted between two consecutive newTruck events.* Asser-tion1 is embedded as a sub-statechart (see Section 2.8.1). Therefore, every time the primary statechart (the TLC) enters the *Red* state, the Assertion1 state is entered, and the sub-statechart is then active, starting its computation in its initial state (*Init* in Figure 4.1b). After-wards, events sensed by the TLC are passed down to the assertion, which then performs state changes like any other statechart. This process continues as long as the TLC is in the *Assertion1* state. Once it leaves the *Red* state it is no longer in the *Assertion1* state, so the assertion statechart is not executing at all. The next time the TLC enters the *Assertion1* state, the assertion sub-statechart starts its computation all over again from its initial state.

The following scenario illustrates the combined behavior of the TLC and *Assertion1*, where, using JUnit terminology (see Chapter 2). TLC is an instance of TLCa, namely the TLC unit under test:

**S4.1.1:**

// startup: the TLC is in state *Yellow*

tLC.start(); // the TLC moves to state *Active/Green*

tLC.timeout(); // the TLC moves to state *Active/Red/<On,Count_0,*
// *Assertion1* (nCnt==0)>

tLC.newTruck(); // the TLC stutters in the same configuration, and
// the assertion moves to A

tLC.newCar(); // the TLC stutters, and the assertion moves to A
// (nCnt==1)

tLC.stopAssertion (); // tLC moves to *Assertion1_off*, leaving
// Assertion1

tLC.timeout(); // tLC moves to *Active/Green*, leaving Assertion1

tLC.timeout(); // the TLC moves to state *Active/Red/<On,Count_0,*
// *Assertion1* (nCnt==0)>

tLC.newTruck(); // the TLC stutters in the same configuration, and
// the assertion moves to A

tLC.newCar(); // the TLC stutters, and the assertion moves to A
// (nCnt==1)

tLC.newCar (); // the TLC stutters, and the assertion moves to A
// (nCnt==2)

tLC.newCar (); // the TLC stutters, and the assertion moves to A
// (nCnt==3)

tLC.newCar (); // the assertion moves to *Error* because *nCnt*>N
// induces a TLC move to the *Green* state via the
// *terminal* state [see Section 4.1.2]

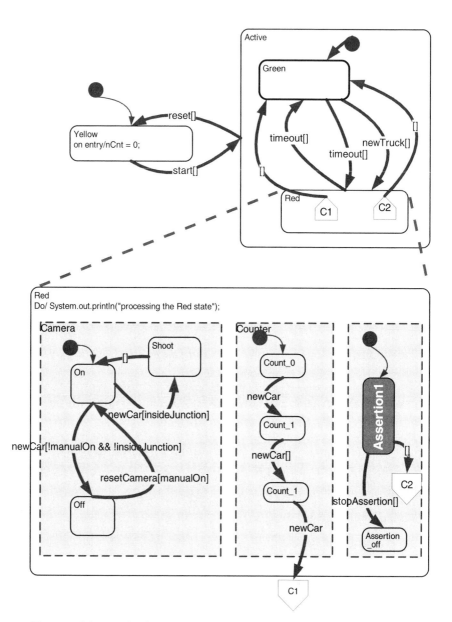

FIGURE 4.1a   TLC with embedded assertion (Assertion 1, Figure 4.1b).

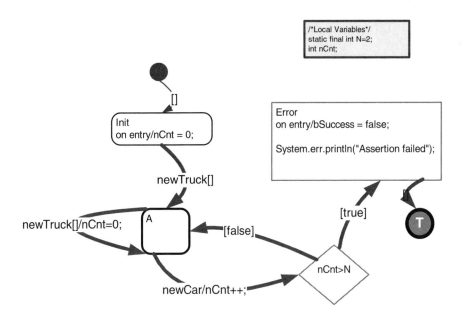

FIGURE 4.1b   Assertion1 statechart.

As its name implies, *Assertion1* is expected to assert about the primary statechart, (the TLC)—i.e., to make a statement about its correctness. It does so using a built-in Boolean variable named *bSuccess*, and a corresponding method called *isSuccess()*, both automatically created by the code generator. The variable *bSuccess* is true by default; therefore, whenever the assertion detects a violation of the requirement it assigns *bSuccess = false*, as in Figure 4.1b. The method *isSuccess()* returns the value of *bSuccess* mainly for the purpose of JUnit testing, which we will discuss in Section 4.1.5, and automatic white-box testing, which we will discuss in Section 5.7.4.

The bottom line is that any sub-statechart with a *bSuccess* Boolean variable and a corresponding *isSuccess()* Boolean method is an assertion statechart.

## 4.1.2. Terminal State and Assertions

Transitions issuing directly from a sub-statechart, assertion or other, are either labeled with an event or are not. Consider first a transition labeled with an event, such as *Assertion1* → *Assertion_off*, labeled with the event *stopAssertion* in Figure 4.1a. It behaves as expected, namely as illustrated in scenario S4.1.1 above: when the *newCar* event is sensed by the TLC, the TLC "jumps out" of the assertion state and therefore stops processing the assertion.

In contrast, an unlabeled transition out of a sub-statechart fires only when the sub-statechart reaches a *terminal* state (labeled with a **T**), as is the case at the end of scenario S4.1.1.

This kind of transition, which fires when the assertion reaches its terminal state, is used for runtime recovery. For example, using scenario S4.1.1 the TLC recovers from an error that was detected by the assertion, and recovery is manifested by the TLC transitioning to the *Green* state.

## 4.1.3. Assertion Actions

Since it is simply a statechart for all practical purposes, an assertion statechart can execute custom actions. One example is the assertion in Figure 4.1b, which logs the fact that the assertion failed. Another is the activation of methods that compile statistics (e.g., how often, on average, does the assertion fail across the entire test suite).

## 4.1.4. Separating Assertions from the Statechart Model: Interfaces and Event Mapping

It is important to separate assertions from the statechart model so that they can be reused (we will discuss reuse in Chapter 5). Java interfaces are a convenient tool to do that. The StateRover uses two interfaces for that purpose: *ITRPrimary* for the primary statechart and *ITRAssertion* for the assertion statechart. Since the interfaces are in source code form, they can be extended to allow a primary to pass custom information down to the assertion and vice versa. An important method in *ITRPrimary* is *getTime( )*, which is also important in *ITRAssertion*. Two other methods in *ITRAssertion* are *isSuccess( )*, which we discussed earlier, and *timeoutFire( )*. Both *getTime( )* and *timeoutFire( )* are used for modeling real-time constraints, which we will discuss in Section 4.1.6.

An assertion statechart doesn't know who its primary statechart is; i.e., it doesn't know its type. The assertion statechart refers to the primary as *primary*, of type *ITRPrimary*. Hence, if an assertion needs to examines the primary statechart's time it calls *primary.getTime*().

Event mapping is the simple mapping of event names in the primary statechart name space to the assertion name space. For example, the assertion in Figure 4.3a uses the generic names *P* and *Q* but the particular application of this assertion, inside the TLC, uses the *newCar* and *newTruck* events. Event mapping would then map *newCar* to *P* and *newTruck* to *Q*.

## 4.1.5. Simulating and Testing Assertions

Figure 4.2 illustrates the JUnit-based testing architecture we use. Tests, which consist of sequences of events and timing information, are either hand coded or automatically created by the white-box test generator described in Chapter 5. A test exercises the primary statechart model, which then automatically exercises the embedded assertion. The assertion feeds back a Boolean success value—that is, an *isSuccess*() value—to the JUnit-based test, which then announces failure or success accordingly.

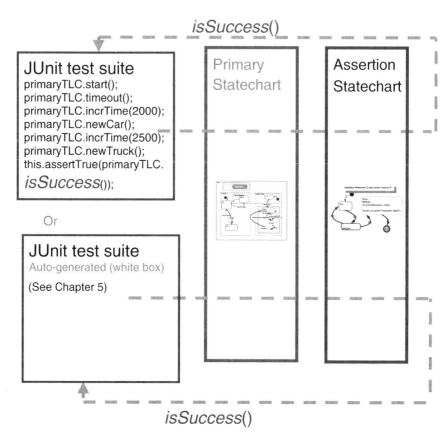

FIGURE 4.2   JUnit-based testing of statecharts and assertions.

Assertion simulation is an important part of the assertion development process, which we will discuss in Chapter 5. Statechart assertion simulation is no more than JUnit testing of that assertion, as illustrated in the following JUnit (Java) code that tests the assertion of Figure 4.1a using the input scenario Q.E.E.E.R:

```
import junit.framework.*;

public class TestAssertion1 extends TestCase {

    private Assertion assertion = null;
```

```
protected void setUp() throws Exception {

    super.setUp();

    assertion = new Assertion();

}


protected void tearDown() throws Exception {

    assertion = null;

    super.tearDown();

}


public void testExecTReventDiapatcher() {

    assertion.Q();

    assertion.E();

    assertion.E();

    assertion.E();

    assertion.R();

    assertTrue(assertion.isSuccess());

}


}
```

## 4.1.6.  Asserting about Time

Real-time constraints exist in most real-life assertions. Consider a generic statechart assertion for the NL requirement:

**R4.1.6**: *A P event is not permitted within the T time interval follow-ing an event Q.*

It is important that the time measurement, whether simulated time or real-time, is the same time measurement that the primary sees.

There are two principal ways of asserting about time in asser-tion statecharts. The first is the *getTime*() method, in which *primary.getTime*() returns a long value of the time reading as seen by the pri-mary; in other words, this is a polling method. The second method is event driven, whereby a built-in event, *timeoutFire*(), fires after a certain amount of time has elapsed.

Figure 4.3a illustrates a statechart assertion for R4.1.6 using *prima-ry.getTime*() polling. Using *primary.getTime*() is simple. First, the asser-tion stores the initial time polled, as done in the Init state. Later it uses that time and compares it with the current *primary.getTime*() value.

Figure 4.3b illustrates a statechart assertion for R4.1.6 using the event-based *timeoutFire*() method. Here, the assertion uses a *TRTime-out* timer object, which is set to the time interval $T$ (in the variables box). Every time the *Init* state is reached, the timer object is reset—i.e., time counting starts from 0. The *timeoutFire* event fires, and conse-quently the transition from $A$ to *Init* if the time interval $T$ has elapsed.

Clearly, as evident from Figures 4.3a and 4.3b, the event-driven approach is more readable than the polling approach.

Both Figures 4.3a and 4.3b use the following informal claim illustrated in Figure 4.3c:

**C4.1.6a**: *for any illegal scenario, if P happens within interval T of some event Q then it must also be within interval T of a more recent event Q* (denoted as Q′ in the figure).

This claim means that the statechart implementations of Figures 4.3a and 4.3b must consider and measure the interval $T$ with respect to *most recent* event $Q$.

The opposite case occurs for a requirement with NL with a lower-bound time constraint, such as:

**R4.1.6b**: *A P event is not permitted after T time of an event Q.*

Figure 4.3d is a statechart assertion for this requirement. It uses the following informal claim illustrated in Figure 4.3e:

**C4.1.6b:** *for any illegal scenario, if P happens after time T of some event Q* (denoted as Q′ in the figure) *then it must also be after time T of an earlier event Q.*

This claim means that the statechart implementations of Figure 4.3d must consider and measure the interval *T* from the *first* event *Q*.

Having to decide whether an assertion under development falls under the auspices of C4.1.6a or C4.1.6b is arguably confusing and counterproductive. For example, requiring *P* to occur rather than barring it from occurring in R4.1.6a and R4.1.6b reverses the decision. You can avoid this kind of confusion by using nondeterminism, as described in Section 4.2.

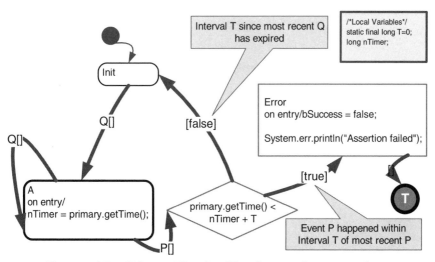

FIGURE 4.3a   Using getTime() polling from a primary statechart, per requirement R4.1.6.

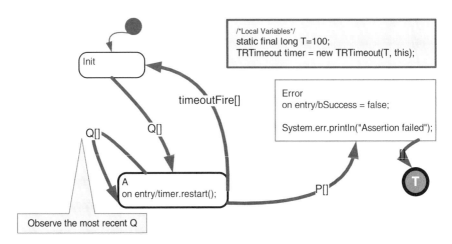

FIGURE 4.3b   Using timeoutFire(), per requirement R4.1.6

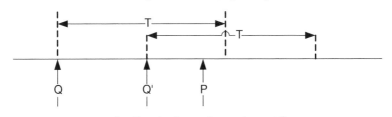

FIGURE 4.3c   An illustration of claim C4.1.6a.

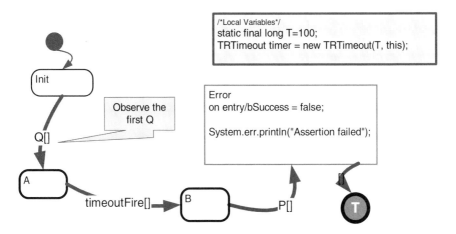

FIGURE 4.3d   A statechart assertion for requirement R4.1.6b.

FIGURE 4.3e   An illustration of claim C4.1.6b

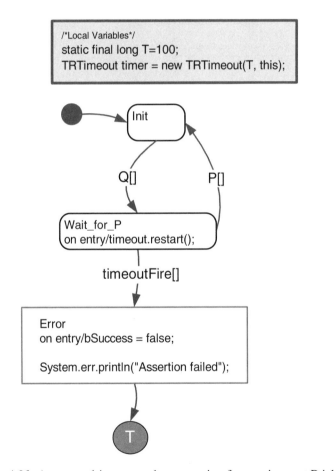

FIGURE 4.3f   An event-driven statechart assertion for requirement R4.1.6.1.

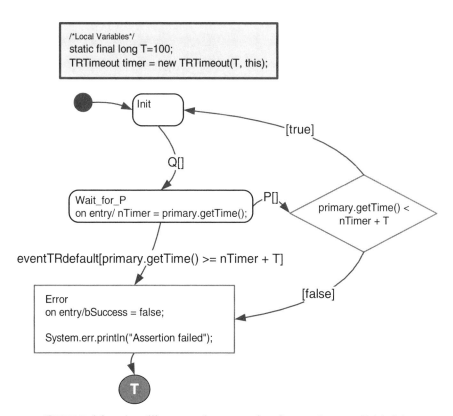

FIGURE 4.3g   A polling statechart assertion for requirement R4.1.6.1.

Testing and simulation time is set in the primary using the *setTime()* and *incrTime()* methods. For example, the following JUnit code executes a TLC scenario with real-time and simulated-time information:

```
tLC.start();

tLC.incrTime(2000); // increments time by 2000 units

tLC.newTruck (); // mapped to Q in the assertion
                 // statechart

tLC.incrTime(2500);

tLC.newCar(); // mapped to P in the assertion statechart

this.assertTrue(tLC.isSuccess());
```

**Bounded Liveness Assertions**

Consider the following *liveness* requirement:

**R4.1.6.1**: *A P event must occur within T time of an event Q.*

Figures 4.3f and 4.3g illustrate the two respective statechart implementations. Say the external entity $E$ that generates the $P$ and $Q$ events dies while the two statechart assertions are in their respective *Wait_for_P* states. Clearly, this scenario should induce assertion failures because event P *does not occur* within time $T$ of event $Q$. However, while waiting in state *Wait_for_P* for event $P$, the event-driven statechart is smart enough to conclude a requirement violation because it creates the *timeoutFire* event locally and does not depend on the dead entity $E$ for doing so. The polling statechart, in contrast, depends on some event to be received from $E$ and only then it checks the timing limit; since $E$ is dead it cannot conclude that the assertion failed.

We say that the polling statechart assertion of Figure 4.3g is monitoring  the liveness requirement in the *weak sense* because the entity $E$ must be live or else the assertion is not accurate.

## 4.1.7. Asserting about States of the Primary

As we will see in Chapter 5, an assertion is expected to assert about a primary statechart mostly from a black-box perspective—namely, with as few references to the internal structure of the primary statechart as possible.

Nevertheless, for those assertions that do require peering into the primary's behavior, the observation must go through the interfaces because the assertion knows the primary only as an *ITRPrimary* object.

The Boolean *isState()* method provides a polling-based method for asserting about the primary state, such as the *isState("Red")* query of the state of the TLC in Figure 4.1a.

The corresponding event-based methods are *primaryEntered()* and *primaryExited()*. For example, Figure 4.4a, which is a variant of the assertion in Figure 4.3b, illustrates the NL requirement for the TLC in Figure 4.1a:

**R4.1.7a**: *the primary is not allowed to enter state Shoot in the T time interval following an event Q.*

Similarly, an assertion can assert about the primary entering and leaving flowchart boxes—for example during a workflow process. The corresponding events are *primaryFlowchartEntered()* and *primaryFlowchartExited()*, as shown in Figure 4.4c, which formally asserts that:

**R4.1.7b**: *the primary statechart in Figure 4.4b always reaches the workflow step named flowB within T time units after entering state Count.*

Note that *primaryFlowchartEntered()* and *primaryFlowchartExited()* induce an atypical event frequency in the assertion. Typically, an event assertion, being an event that is sensed by the primary and passed down to the assertion, fires during the primary's major cycles, namely when the primary attempts to jump from one state to the next. These events, however, fire during the primary's minor cycle, while the primary is processing a flowchart, which happens within a single major cycle. In Figure 4.4b, for example, say the primary is in the *Count* state. After a *newCar* event is sensed, the primary statechart will traverse the flowchart all the way, through flowchart activity boxes *flowA* and *flowB*, back to the *Count* resting state along the way, inducing four events for the assertion to witness: *primaryFlowchartEntered*("flowA"), *primaryFlowchartExited*("flowA"), *primaryFlowchartEntered* ("flowB"), and *primaryFlowchartExited*("flowB").

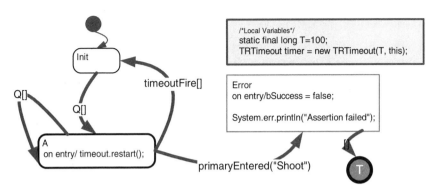

a. An assertion using *primaryEntered*, per requirement R4.1.7a

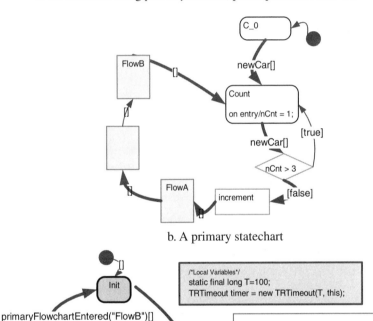

b. A primary statechart

c. An assertion about the primary of Figure 4.4b
using primaryFlowchartEntered

FIGURES 4.4a–c.

## 4.1.8. Scoping Assertions

Statechart assertions are naturally scoped by the context of their sub-statechart. For example, *Assertion1* in Figure 4.1a is active only while the TLC is in the *Red* state. Whenever the TLC exits the *Red* state (e.g., when a *timeout* event triggers a transition to the *Green* state) it is no longer in state *Assertion1* and consequently the assertion is no longer performing any computation. Scoping allows an assertion to be active precisely where we need it, instead of having to be global.

## 4.1.9. Nesting Assertions

Consider again NL requirement R4.1.6. Say that P is not a basic proposition but rather the equivalent of the NL requirement:

**P***: more than N newCar events detected.*

When nested, the composite requirement is similar to:

**R4.1.9**: *No more than N newCar events are allowed within the T time interval following an event Q.*

Figure 4.5 illustrates an assertion for this NL requirement that uses assertion nesting within the TLC in Figure 4.1a. *Assertion1* implements R4.1.6 with a generic sub-requirement, R.4.1.9 (instead of proposition P), implemented by *Assertion2*.

When the TLC is in the *Assertion1* state, all events that it senses, including the *newCar* event, are passed down to the *Assertion1* statechart. *Assertion1* highlight the use of event mapping (see Section 4.1.4) having the primary's *newCar* event mapped to its *Q* event. When the *Assertion1* statechart is in the *Assertion2* state, it passes its events to the *Assertion2* statechart, which in turn maps its own primary's *Q* event to its own *newCar* event.

When the *Assertion2* statechart detects an error—i.e., when it detects more than *N newCar* events—it indicates a failure (*bSuccess*

= false), and moves to the *terminal* state. Consequently, *Assertion1* moves, via the unnamed transition, to the *Error* action box, flags an error, and moves to its terminal state. The primary then recovers by transitioning via the "C2" connector to the *Green* state (on another, invisible, diagram page).

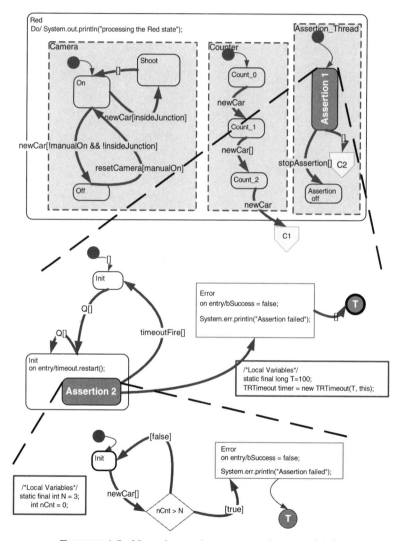

FIGURE 4.5   Nested assertions per requirement 4.1.9.

## 4.1.10.  Chaining Assertions

Again, consider NL requirement R4.1.6. Say that when it fails (i.e., *P is detected within the time interval T after Q*), we would like to verify that neither *P* nor *Q* is detected within the subsequent interval T1. The aggregate NL requirement is thus:

**R4.1.10**: *If  P is detected within the interval T after Q, then neither P nor Q is detected in the subsequent T1 interval.*

The chained assertion in Figure 4.6 implements R4.1.10 using assertion chaining. When *Assertion1* detects a *P* within interval *T* of a *Q*, it moves to its terminal state and the primary TLC statechart transitions to the *Assertion2* state and launches the second part of the composite assertion.

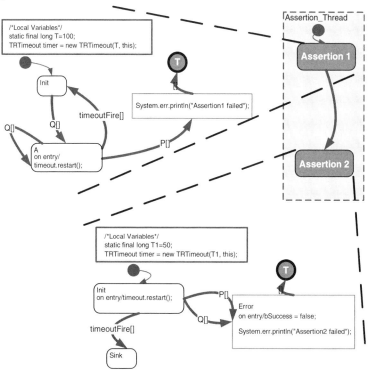

FIGURE 4.6   Chained assertions per requirement 4.1.10.

## 4.1.11. Default Events

The default event (*eventTRdefault* using the StateRover syntax) fires when the assertion statechart receives an event that induces no state change in the statechart. In other words, the default event is to a statechart as an *else* is to an *if-then-else* conditional block of code.

Figure 4.7a illustrates an assertion with a default event. It implements the following NL requirement:

**R.4.1.11**: *No event other than P is allowed between an event Q and a subsequent event R.*

Now consider the following scenario written using JUnit syntax:

```
assertion.Q(); // move to state A

assertion.P(); // move from A to A (i.e., eventTRdefault
               // does not fire)

assertion.Q(); // error, indeed eventTRdefault fires
               // and the assertion flags an error

assertFalse(assertion.isSuccess());
```

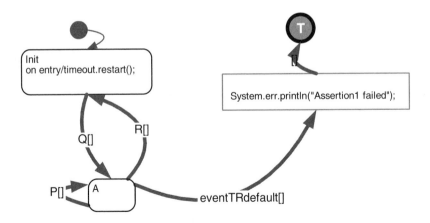

FIGURE 4.7a  The default event: eventTRdefault per requirement 4.1.11.

## 4.1.12. Assertions as Flag Raisers

As we will see in Chapter 5, assertions are usually used to flag errors. Indeed, in the examples so far, an assertion transitions to its terminal state when an error is detected. Stated differently, in the examples an assertion transitions to its terminal state whenever *bSuccess* = false.

Sometimes, however, we might want to flag sequences we intuitively feel are *good*, as in the case of nested assertions using a positively specified sub-assertion. To that end, we need only to *reverse* the way we used the *bSuccess* flag: to set it to *false* in the initial state and to *true* whenever a desired sequence is caught.

In the final analysis, a statechart assertion doesn't really "prefer" one kind of behavior (good or bad) more than the other. All it really does is raise a flag. It does so by reaching a terminal state as a result of certain input sequences, and possibly assigning *bSuccess* along the way. It is up to the user of the assertion (e.g., the primary statechart or the JUnit test suite) to interpret those manifested outcomes.

## 4.1.13. Inverting Deterministic Statechart Assertions

Consider the following requirement, the inverse of R4.1.6.1 and associated statechart assertion of Figure 4.3f:

**R4.1.13**: *there is some event Q in the future for which event P does not occur in the subsequent interval T.*

The statechart assertion of Figure 4.7b realizes this requirement. Note that it is almost identical to Figure 4.3f. In fact, the most significant change is the inversion of the *bSuccess* logic; is now *false* by default and set to *true* in the bottom state. This is exactly the same as we did in Chapter 1 when we inverted a DFA.

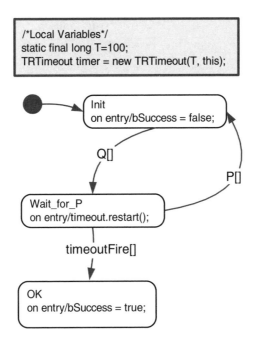

FIGURE 4.7b   Inverting a deterministic assertion.
Statechart assertion for R.4.1.13.

## 4.2. Nondeterministic Statechart Assertions

We discussed nondeterminism in Chapter 1, describing deterministic finite automata (DFA) and nondeterministic finite automata (NFA). However, there are few, if any, references to nondeterminism in the context of statecharts or FSMs. In fact, in Section 2.4.1, we declared nondeterminism (also called a race condition) illegal. That is the case for ordinary statechart models; assertion statecharts, however, can benefit from nondeterminism: we can use it to write readable and simple assertions for complex requirements. Note, however, that nondeterminism is useful only when used within a statechart assertion, and cannot be used within a statechart model.

Figure 4.1b shows a deterministic statechart (assertion). By deterministic we mean that in every cycle, given the input event and conditions, the statechart has one state configuration it decides to jump to. In other words, a deterministic statechart has a linear sequence of configurations. Accordingly, the statechart in the figure (which is also a flat FSM, having no hierarchy or concurrence) boots-up in the *Init* state. Then, after receiving the *newTruck* event it jumps to state *A*; there is no other option. Next, after receiving the *newCar* event it returns, via the visual switch, to state *A*; again, there is no other option. And so on.

In contrast, Figure 4.8a illustrates a nondeterministic statechart assertion for the NL requirement:

**R4.2**: *No more than CNT newCar events can be sensed within any single T second interval.*

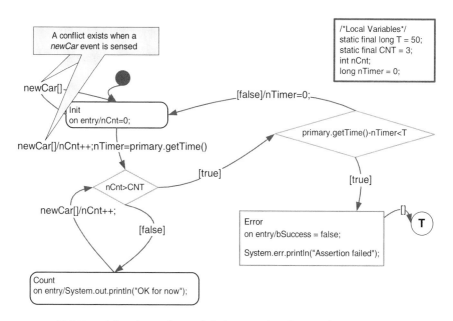

FIGURE 4.8a   A nondeterministic assertion for requirement R4.2.

As indicated in the figure, there is an apparent conflict when a *newCar* event is sensed while the statechart is in the *Init* state: is *Init* the next state or should the computation flow through the visual switch? Under a deterministic statechart interpretation, this is a race condition and therefore illegal. Indeed, the plain vanilla StateRover code generator will issue an error message. However, the assertion is a perfectly good *nondeterministic* statechart.

This example illustrates the nature of nondeterminism in statecharts: if, during any cycle, there is a *conflict* in the next state configuration assignment, the statechart is nondeterministic. In that case, therefore, we must use a special code generator, one that implements nondeterminism.

## 4.2.1. Object-Oriented Informal Semantics for Nondeterministic Statecharts

Let's see how the nondeterministic statechart in Figure 4.8a behaves (assuming T = 50 and CNT = 3), using the following JUnit/Java-based scenario:

### S4.2.1:

```
1. assertion. incrTime (25);

2. assertion.newCar(); // time now is 25

3. assertion.incrTime(50);

4. assertion.newCar(); // time now is 75

5. assertion.incrTime(20);

6. assertion.newCar(); // time now is 95

7. assertion.incrTime(10);

8. assertion.newCar(); // time now is 105
```

Clearly, the scenario should induce an assertion fail according to R4.2 because more than three cars were detected within the 50-second interval starting at time 70 for example.

The statechart boots up in the *Init* state. At this point there is one state-configuration, which consists of *Init*. After the first *newCar* event (line #2), because of the nest state assignment conflict, the nondeterministic statechart forks this state-configuration into two state-configuration objects; one, denoted C1, is in the *Init* state with *nCnt==0*; the other, denoted C2, goes via the decision switch to the *Count* state, with *nCnt==1*. Each state-configuration behaves as if it were the only one that exists—namely, as it would under deterministic circumstances if there were no conflict at all.

Let's record all the pieces of information that form these two state-configuration objects: for C1, <state==*Init*, *nCnt*==0, *nTimer*==0, *bSuccess*==false>, and for C2, <state==*Count*, *nCnt*==1, *nTimer*==25, *bSuccess*==false>.

After the second *newCar* event, (line #4) state-configuration object C1, being in the *Init* state, faces a conflict again and therefore forks into two state-configuration objects: C11 in the *Init* state, with *nCnt*==0, and C12 in the *Count* state, with *nCnt*==1. Meanwhile, state-configuration object *C2* loops via the decision switch to the *Count* state, with *nCnt*==2. Note that C2 and C12 are *different* state-configuration objects, although both are in the *Count* state: C2 sees *nCnt*==2, whereas C12 sees *nCnt*==1. It is not as if one state-configuration object is the "correct" one and the other is not. Rather, they are distinct, each having a distinct value for the data member variable *nCnt*. Similarly, they have different values for their *nTimer* member variable, as we will see below. The value of the *bSuccess* member variable, however, is *true* in all state-configuration objects. Specifically, the three state-configuration objects are now (C1 is substituted by C11 and C12):

- C11, which is *<state==Init, nCnt==0, nTimer==0, bSuccess==*false>

- C12, which is <state==*Count, nCnt==1, nTimer==*75, *bSuccess==*false>

- C2 , which is <state==*Count, nCnt==2, nTimer==*25, *bSuccess==*false>

Hence, state-configuration objects are distinguishable by their constituent states as well as by the values assigned to data members. In contrast, in NFA formalism, absent the notion of variables, state-configuration objects differ only in their state component.

After the third *newCar* event, (line #6) state-configuration object C11 forks into two state-configurations: C111 is <state==*Init, nCnt==0, nTimer==0, bSuccess==*false>, and C112 is <state==*Count, nCnt==1, nTimer==95, bSuccess==*false>. State-configuration object C12 is now <state==*Count, nCnt==2, nTimer==75, bSuccess==*false>, and state-configuration object C2 is now <state==*Count, nCnt==3, nTimer==25, bSuccess==*false>.

After the fourth *newCar* event, (line #8) there are five state-configuration objects:

- <state==*Init, nCnt==0, nTimer==0, bSuccess==*false>,

- <state==*Count, nCnt==1, nTimer==105, bSuccess==*false>,

- <state==*Count, nCnt==2, nTimer==95, bSuccess==*false>,

- <state==*Count, nCnt==3, nTimer==75, bSuccess==*false>, and

- <state==*Terminal, nCnt==4, nTimer==25, bSuccess==*true>

Essentially, then, nondeterministic behavior is an extension, or relaxation, of the rules governing deterministic statechart behavior.

Whereas state-configurations in a deterministic statechart progress linearly cycle by cycle, from one uniquely defined configuration object to another uniquely defined next configuration object, in non-deterministic statecharts, they fork each cycle. Each state-configuration object behaves as if it is the only one in the statechart, unaware of the existence of other configuration objects, and when faced with a conflict it tries all possible avenues (e.g., trying, in Figure 4.8a, from a configuration object that is in the *Init* state both the transition *Init→Init* and the transition *Init→Count*) and morphs, or forks, into the resulting configurations, one for each avenue. Clearly deterministic behavior is a private case of nondeterministic behavior, where absent conflicts the forking degree is always one.

**Nondeterministic Statechart Actions**

Let's look at actions performed by nondeterministic assertions. For example, the *Count* state in Figure 4.8a has an embedded Java action. This action is performed by *every* configuration object that reaches that state. Hence, in the example above, after four *newCar* events there are three state-configuration objects in *Count*; each has its own copy of the member variables of the assertion-statechart class, and each performs actions, potentially using its own copy of the variables, as if it were the only state-configuration. In Section 4.2.6, we will examine a mechanism for power users that provides more control over variables and actions.

**The Existential Behavior of Nondeterministic Statecharts**

The two main roles of statechart assertions in general, and nondeterministic assertions in particular, are to announce success or failure via the *isSuccess*() method and then, possibly, to transition to the terminal state so that the primary statechart can perform a recovery.

Because assertions are usually used to flag errors, *isSuccess*(), which monitors the *bSuccess* variable, is *true* by default. It is our responsibility to set it to *false* when the assertion fails, as done in the *Error* activity box in Figure 4.8a.

Because *isSuccess*() flags an error if one state-configuration exists that detects an error (assigns *bSuccess*=false), it is called *existential*.

Likewise, terminal state behavior is existential: if at least one state-configuration exists that is in a terminal state, the nondeterministic statechart assertion wrapper considers itself to be in a terminal state.

An alternative to existential *isSuccess*() and terminal state behavior is *universal* (*forall*) similar to the behavior exhibited by $\forall$-FA described in Chapter 1. For example, say we want a nondeterministic assertion to fail if and only if *all* state-configurations detect a failure. This kind of behavior is achievable using shared variables, described in Section 4.2.6.

## 4.2.2. Code Generation and Performance Issues

The semantics discussed earlier are also relevant to the way the StateRover code generator implements nondeterministic assertions. As with deterministic statecharts and statechart assertions, a nondeterministic statechart is realized as a class, as illustrated in Figure 4.8c. Note, however, that it can be used only as an assertion.

As illustrated, the nondeterministic statechart class encapsulates and manages state-configurations internally, using an inner *Config* class. Variables and methods declared inside the assertion statechart diagram (e.g., *nCnt* in Figure 4.8a) are considered local to a *Config* inner object. However, at the end of the day, the assertion needs to announce a success or failure decision. That is done by the *isSuccess*() method, which makes an *existential* decision: if *any Config* object has *bSuccess*==false then and only then *isSuccess*() returns false. Similarly, the nondeterministic assertion statechart announces that it has reached a terminal state if *any Config* object has reached a terminal state.

Though the *isSuccess()* and terminal state decisions seem global, they are not. Rather, they are shared inside a statechart assertion object. If the same assertion statechart is instantiated more than once, each assertion statechart object has its own internal Configs and each manages its own existential *isSuccess()* and terminal states.

Based on our discussion so far, it seems as if the number of state-configurations is nondecreasing. However, it is quite possible that the same configuration will be generated more than once within some cycle. In such cases there is no need to preserve all identical state-configurations, because their future behavior depends only on the data stored inside the configuration and the inputs, and those are identical, resulting in identical future behavior. For example, consider the again statechart assertion in Figure 4.8a (again T=50, CNT=3) under the following scenario:

**S4.2.2a:**

```
1. assertion.newCar(); // time is 0

2. assertion.incrTime(125);

2. assertion.newCar(); // time now is 125

3. assertion.incrTime(50);

4. assertion.newCar(); // time now is 175

5. assertion.incrTime(20);

6. assertion.newCar(); // time now is 195
```

After the fourth *newCar* event (time 195) there are two state-configurations in the *Init* state: one resulting from the transition *Init→ Init*; the other resulting from a transition from the *Count* state, via the two visual switches, and back to *Init* (the last switch checked primary. getTime()–nTimer<T, i.e., 195–0 < 50, which is false). The two are identical, being <state==*Init*, *nCnt*==0, *nTimer*==0, *bSuccess*==false>. Therefore, they are now considered one configuration.

Now consider the assertion statechart in Figure 4.8b, a variant of the one in Figure 4.8a. It has the same behavior as the original: namely, it returns the same *isSuccess()* value given the same input sequence. The difference between the two versions is that the first one never has more than five *Config* objects at any given time: one in the *Init* state, three in the *Count* state (with *nCnt* =1, 2, and 3), and one in the terminal state. The modified version, on the other hand, can have many more, as in the following scenario:

**S4.2.2b:**

```
1.  assertion.newCar();  // time is 0

2.  assertion.incrTime(200);

2.  assertion.newCar();  // time now is 200

3.  assertion.incrTime(200);

4.  assertion.newCar();  // time now is 400

5.  assertion.incrTime(200);

6.  assertion.newCar();  // time now is 600

7.  assertion.incrTime(200);

8.  assertion.newCar();  // time now is 800

9.  assertion.incrTime(200);

10. assertion.newCar();  // time now is 1000
```

This scenario induces seven *Config* objects: one in the *Init* state, three in the *Count* state (with *nCnt* = 1, 2, and 3), and three in the *Sink* state (distinguishable by their *nTimer* values). In fact, one can easily design an assertion in which the number of *Config* objects grows in an unbounded manner.

Such potentially unbounded behavior is obviously undesirable during runtime. We will discuss techniques to tackle this problem in Section 4.5.1.

Note that a simple *Sink* state on-entry action, *nTimer=0*, solves the problem for Figure 4.8b

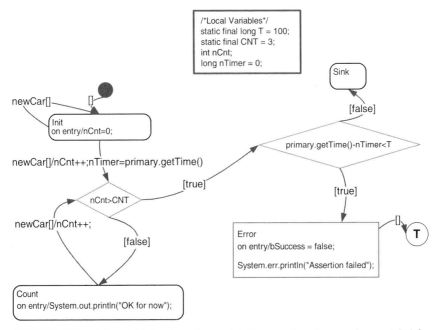

FIGURE 4.8b   A less efficient nondeterministic assertion for requirement R4.2.

```
class Assertion {
  private TRPSSet configSet;

  Assertion() {
     eventTRreset();
  } // constructor

  Boolean isSuccess() {
    return configSet.isSuccess(); // an existential decision
  }

  /* Event handler for event:newCar()*/
  int newCar(){
    configSet.newCar();
  } //newCar()


  //---------- inner classes --------------------------
  class TRPSSet {
     private TreeSet mySet; // set of Config's
     private WinnerPSConfigs winnerPSConfigs; // winner Configs

     boolean isSuccess() {
       ...// an existential decision amongst Config's in mySet
     }
     ...
  } // inner class TRPSSet

  // a Config class
  class TRNSConfig implements Cloneable {
     ...
     /* code similar to a deterministic statechart implementation */
     ...
  } // inner class TRNSConfig

} // class Assertion
```

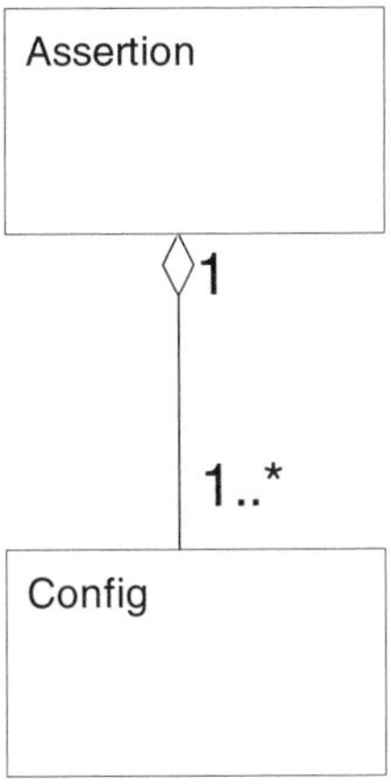

FIGURE 4.8c.

## 4.2.3. Modeling Past Time

Nondeterminism is useful for modeling requirements that are anchored in the past. Consider the NL requirement:

**R4.2.3**: *At most three cars should be detected within the 30-second interval that precedes a time in which the Primary statechart enters the Green state.*

The nondeterministic statechart assertion in Figure 4.9 realizes this requirement. It nondeterministically chooses (or "guesses," as discussed in Section 1.5.2) the event in the sequence of *newCar* events that violates the requirement and records the arrival time of that event. When the primary enters the *Green* state, the assertion verifies that more than three *newCar* events were received and no more than 30 seconds has elapsed since the recorded time.

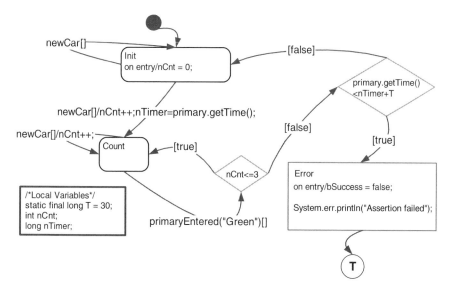

FIGURE 4.9    Nondeterminism used for a past-time requirement.

## 4.2.4.  Default Events for Nondeterministic Assertions

Recall the built-in event *eventTRdefault* which is to statecharts as an "else" is for an if-then-else conditional Java statement. The *event-TRany* event is somewhat similar to *eventTRdefault*. It fires if *any* event fires during that cycle. We will illustrate this type of event in Section 4.2.10.

## 4.2.5. Power Usage: Prioritization

Prioritization is a power-user technique for limiting the scope of the existential behavior described earlier. The idea is simple: whereas existential behavior considers any state-configuration as a candidate for catching a violation, prioritization limits this set to a set of state-configurations called *winners*. If a state configuration that detects an assertion failure (i.e., *bSuccess*=false) is not a winner, it doesn't count; namely, it won't cause the overall nondeterministic assertion to fail.

Prioritization works in the following way. First, each state in the assertion statechart has a positive integer priority, with 1 the default priority. As the assertion code executes, cycle by cycle, state-configurations visit various states, each with its own priority. A state-configuration is then said to have a priority, which is simply the highest priority of all the states it visits (if the statechart is flat and has no orthogonality, it's visiting only one state). Hence, every cycle, some state-configurations potentially have higher priorities than others. A state-configuration whose priority is *not* lower than any other state-configuration is called a *winner* configuration, and the existential rule is applied only to winner configurations; namely, *isSuccess*() is false if and only if *some* winner-configuration has *bSuccess*==false. Likewise, the assertion's terminal state is considered to have been reached if and only if *some* winner-configuration has reached the terminal state. Note that when no priorities are used—i.e., when all states have the same priority—every state-configuration is a winner.

Consider the NL requirement:

**R4.2.5**: *If, between start and end events, four or more cars are detected in any interval [A,B], then three or more trucks must be detected in some interval [C,D].*

In the spirit of assertion reuse, which we will discuss in Chapter 5, say that we reuse two assertions: (i) for the sub-requirement *four or more cars are detected in some interval [A,B]*, and (ii) for the sub-requirement *three or more trucks are detected in some interval [C,D]*. Let's denote these assertions A1 and A2. Using them we can rewrite R.4.2.5 as:

**R4.2.5a:** *If, between start and end events, A1 raises a flag, then A2 must also raise a flag.*

Figure 4.10 illustrates the assertion statechart for requirement R.4.2.5 using these sub-assertions. This statechart is nondeterministic, since the choice between the lower and upper paths is non-deterministic. When A1 raises a flag, the lower path is traversed to *Priority_1* activity box, and when A2 raises a flag then the upper path is traversed to *Priority_2* state, the only state with priority greater than 1. Note how, along the lines of our discussion in Section 4.1.12, we describe assertion A1 and A2 as raising flags and not as succeeding or failing.

Now let's examine the assertion's behavior using two scenarios.

**S4.2.5a**: *In the period between start and end events: four (i.e., too many) cars are detected in some interval [A,B] and one truck is detected throughout the entire period.*

In this scenario (Figure 4.10b), assertion A1 raises a flag, moves to its terminal state, and therefore induces a transition to the *Priority_1* activity box. Assertion A2, on the other hand, doesn't raise a flag. Hence, for the assertion, the *Priority_2* state is not reached and therefore all states and all state-configurations have priority 1. Consequently, the action *bSuccess*=false is executed in *Priority_1* and the terminal state is reached by that state-configuration. The method *isSuccess()* then returns false because some winner state configuration has *bSuccess*==false.

**S4.2.5b**: *In the period between start and end events: four (i.e., too many) cars are detected in some interval [A,B] and three trucks are detected in some interval [C,D].*

In this case (Figure 4.10c) both A1 and A2 raise a flag. Hence both *Priority_1* and *Priority_2* are reached. *Priority_1* assigns *bSuccess*=false in the lower state-configuration object, and *Priority_2* assigns *bSuccess*=true in the upper state-configuration object. The winner state-configuration is the upper one, therefore *isSuccess()* takes as its value the upper state-configuration's *bSuccess* value, namely *isSuccess()==true*. Likewise, the overall assertion doesn't reach a terminal state because no winner state-configuration did.

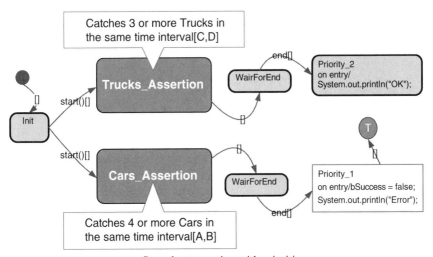

a. Statechart assertion with priorities

FIGURE 4.10a   Using priorities.

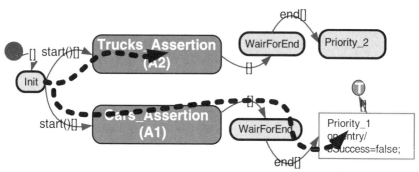

b. The behavior for scenario S4.2.5a

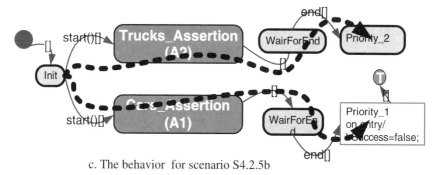

c. The behavior for scenario S4.2.5b

FIGURES 4.10b–c   Using priorities.

A devil's advocate would argue that because R4.2.5 is merely a logical implication there must be a simpler way to formalize it. Indeed, in Section 4.3 we show a simple pattern for logical implication. In R4.2.5, however, the relationships between the intervals [A,B] and [C,D] are unknown; for example, we cannot assume that B < C. It is for this reason that we need to evaluate both paths simultaneously and then implement the logical implication using priorities.

### 4.2.6.  Power Usage: Shared Variables

We've seen that local variables within nondeterministic assertions are local to state-configuration objects. Hence, in Figure 4.9, there is an *nCnt* and an *nTimer* for each state-configuration.

In Figure 4.11, however, *nCnt* is marked as shared. Consequently, all state-configuration objects within a single assertion object refer to the same *nCnt*.

The assertion realizes the NL requirement.

**R4.2.6**: *no more than M times it is permitted that N or more events E occur within time T of event Q.*

The assertion uses the shared *nCnt* to count the number of times *N an E event occurs within time T of event Q.*

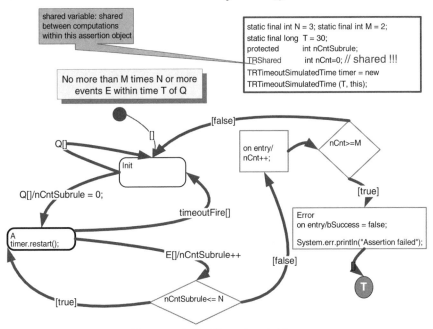

FIGURE 4.11   Shared variables.

## 4.2.7. Power Usage: Orthogonality within Nondeterministic Assertions

Almost all the assertions we have considered thus far, deterministic and nondeterministic alike, didn't use an important statechart feature, orthogonality. Statechart assertions, including nondeterministic assertions, can enjoy all three statechart features: hierarchy, orthogonality, and history.

Consider, for example, a statechart assertion $A_{or}$ that reuses two other statechart assertions, $A_1$ and $A_2$, in a manner that resembles a logical or, namely, $A_{or}$ fails if either $A_1$ or $A_2$ fail. Similarly, consider a statechart assertion $A_{and}$ that fails if both $A_1$ and $A_2$ fail. In Section 4.3 we show how to construct such statechart assertion equivalents of logical *or* and logical *and* of primitive and nested assertions utilizing statechart orthogonality. If either $A_1$ or $A_2$ are nondeterministic, then $A_{or}$ and $A_{and}$ are nondeterministic but both use statechart orthogonality.

## 4.2.8.  Quantification and Nondeterministic Statecharts

Consider a system with a collection of *Car* objects. Now consider the NL requirement:

**R4.2.8a**: *for all Car objects, once event P is sensed by a car then eventually within time T, event Q should be sensed by the same car.*

It is in effect the application of the MTL assertion, *Always P Implies Eventually* $_{\leq T}$ Q, to every *Car* object. This requirement is called a universal quantification ($\forall$, *forall*) over a set of *Cars*. One way to realize R4.2.8a as an assertion is to create a statechart assertion for R4.2.8a, embed it in the statechart for a *Car*, and have all resulting assertion objects (one per car object) coordinate somehow to verify that they *all* succeed when necessary. This approach suffers from two main problems: (i) *Car* is not necessarily a statechart-based controller, so embedding a statechart assertion in it requires some extra work, and (ii) the coordination between the assertions must be coded by hand.

Figure 4.12 shows a different solution using a nondeterministic statechart assertion. Here, a *Car* objects is passed in as an event argument. For example, if event *P* is sensed by car #3, the assertion receives it with car #3 as an argument. Nondeterminism is used to "guess" the violating *Car* object. The *Car* object reference is then stored in the *myCar* local variable. It is used later, if a *Q* event is received, to verify that the *Car* object is the same as the one associated with the *P* event.

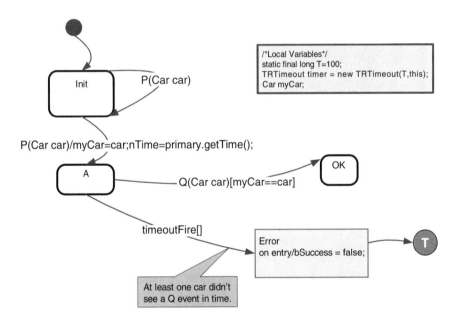

FIGURE 4.12a   An assertion that performs universal quantification.

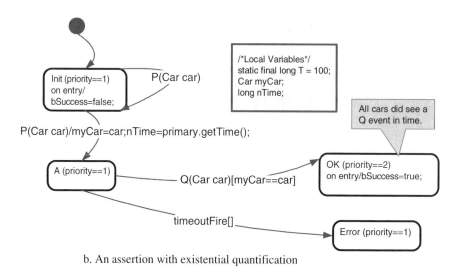

b. An assertion with existential quantification

FIGURE 4.12b   Statechart assertions for quantified requirements.

Let's examine the dual requirement with existential (∃, there exists) quantification.

**R4.2.8b**: *there exists a Car object such that once event P is sensed by this car then eventually within time T event Q should be sensed by the same Car.*

A slight variation to the nondeterministic statechart assertion of Figure 4.12a implements this requirement, as illustrated in Figure 4.12b. The statechart assertion assigns priorities as follows: the *OK* state has priority 2 (the highest) and all other states have priority 1. In addition, the *bSuccess* default value is false and the only state that changes that value is the *OK* state. Now, any computation that reaches *OK* is by definition a winner computation; it has *bSuccess* as true. In this case, even if another computation reaches another state where *bSuccess* is false (e.g., *A* or *Error*), the assertion as a whole will only look at the *bSuccess* value of the winner computation, which is true. On the other hand, if no computation reaches the *OK* state, then the assertion's *bSuccess* value is false.

## 4.2.9. On Deterministic vs. Nondeterministic Statechart Assertions

Clearly, using nondeterminism in statechart assertions doesn't come without a price. The semantics are more complex; fixing a buggy assertion could be harder than in the deterministic case; and, as will be discussed in the sequel, there is a performance penalty.

Nevertheless, in some situations, nondeterminism is useful. The following are some types of requirements that can benefit from non-determinism:

1. *Bounded counting assertions.* For example, the NL requirement:

   **R4.2.9.1:** *No more than N (N > 0) events E within any one minute time interval.*

   Figure 4.13 describes a deterministic statechart assertion for this requirement. It uses a Java data structure (a queue) to record the arrival time of each $E$ event. Once more than $N$ $E$s are detected it compares the difference between the arrival times of the first $E$ and the $N$th, and it flags an error if the difference is less than a minute.

   R4.2.9.1 is a rewrite of requirement R4.2 ($E$ is a *newCar* event); Figure 4.8 is a nondeterministic assertion for R4.2. It is more visual than its deterministic counterpart in that it doesn't use any textual Java facilities. Rather, it guesses the violating $E$ (*newCar*) and then validates that it is a violating $E$.

   The point illustrated by this pair of statecharts is that, given enough time and money, we can write deterministic versions of most, perhaps all, nondeterministic statechart assertions. In fact, with more time, money, and a glimpse at immortality, we can also write them in assembly... The real question to be asked is: *Why bother?* A primary reason for investing in a deterministic version is performance—for example, if the assertion is expected to be active in runtime. Often, however, it is more important to develop a readable, modifiable assertion quickly. To that end, many users prefer the nondeterministic version.

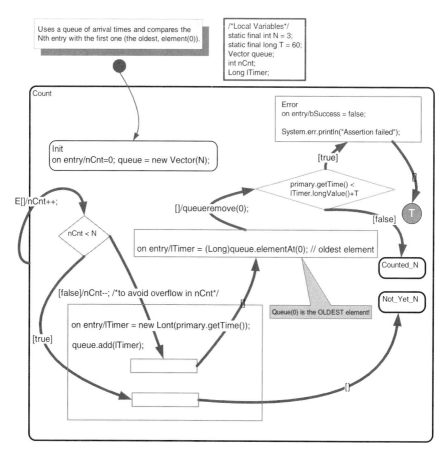

FIGURE 4.13    A deterministic assertion for requirement R4.2.9.1.

2. *Nested requirements.* R4.2.6 is an example of a nested requirement. Figure 4.14 illustrates a scenario that demonstrates the benefit of nondeterminism when used to realize

this requirement. The scenario consists of two $Q$ events and their corresponding $T$ intervals. Note how the third $E$ event is shared by both intervals and needs to be counted by both. A deterministic solution therefore cannot assume that an $E$ event should be "counted-for" or "associated-with" one particular $Q$ event as we did in Section 4.1.6.

The bottom line is that nesting within a requirement tends to make a deterministic solution difficult to write and less readable.

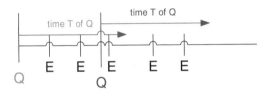

FIGURE 4.14   A scenario illustrating a difficulty associated as a deterministic statechart assertion with realizing requirement 4.2.6.

Figure 4.11 is a (nondeterministic) statechart assertion that realizes R4.2.6.

3. *Nonmonotonic requirements.* First, let's look at a monotonic requirement.

   **R4.2.9.3a:** *No more than N (N > 0) events E between event Q and event R.*

   Figure 4.15a is a deterministic assertion that realizes this requirement. It uses the fact (discussed in Section 4.1.6) that the measurement function (the number of $E$s) is monotonically increasing. Therefore, if there are two or more $Q$ events prior to an $R$ event, the only one that matters is the first one, because it "sees" more $E$s than subsequent $Q$s. For that reason, the assertion in Figure 4.15a looks only at the first $Q$ for every $R$. Similarly, for a variant of R4.2.9.3a that requires *no* fewer *than N events E between event Q and event R*, the only event *Q* that matters is the last one because it "sees" the least amount of *E* events.

In contrast, the following NL requirement is nonmonotonic; it doesn't simply count—it looks at an average.

**R4.2.9.3b:** *For every event Q let nBaseTime be the arrival time (primary.getTime() ) of that event. For every event Q, the average of primary.getTime()–nBaseTime values per event E that occurs within 100 time units of Q is never greater than AVG.*

Figure 4.15b shows a nondeterministic assertion that realizes this requirement.

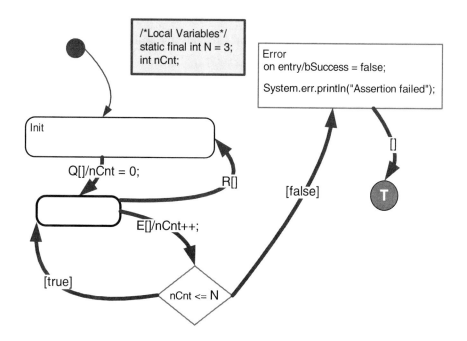

FIGURE 4.15a   A deterministic assertion for requirement R4.2.9.3a.

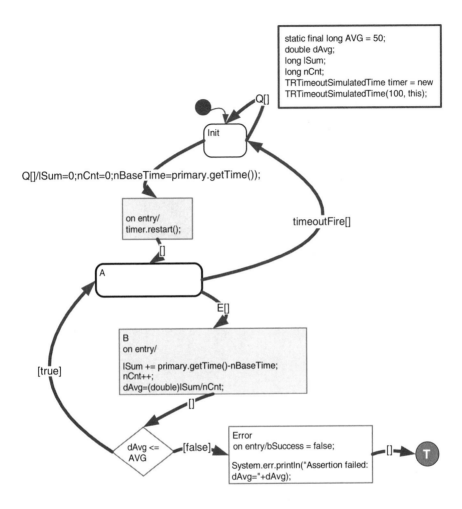

FIGURE 4.15b  A nondeterministic assetion for requirement R4.2.9.3b.

4. *Requirements with both lower and upper time bounds.* First, consider requirements R4.1.6 and R4.1.6b. They contain an upper time bound constraint or a lower bound constraint, but not both. These requirements are realized by the deterministic statecharts of Figures 4.3b and 4.3d.

Now consider the following NL requirement, which has both lower and upper time bounds.

**R4.2.9.4:** *Whenever event Q occurs then P is not allowed outside time interval [A,B].*

Having both an upper and a lower bound, an assertion cannot simply record just the earliest or the latest $Q$. Rather, it must consider every $Q$ event and measure a corresponding time interval [A,B]. A deterministic assertion can do that using a dynamic Java data structure—dynamic because the number of $Q$s within any given interval is unknown a priori. In contrast, the nondeterministic statechart in Figure 4.16 captures the requirement visually.

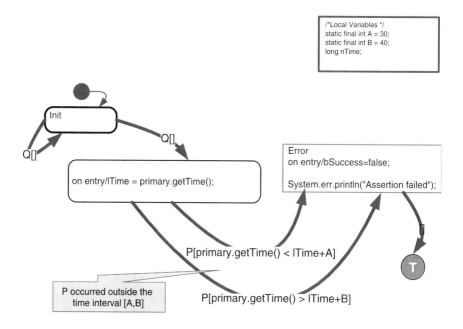

FIGURE 4.16   A deterministic statechart assetion for requirement R4.2.9.4 with both lower and upper time bounds.

5. *Requirements with quantification.* See Section 4.2.8.

6. *Past time requirements.* See Section 4.2.3.

7. *Conditioned requirements (if then else), using priorities.* See Section 4.2.5.

## 4.2.10. Making Nondeterministic Assertions Deterministic

There are several reasons we might want to make an assertion deterministic. For example:

1. Efficiency, particularly for runtime recovery applications, which we will discuss in Section 4.5.

2. Readability. Some developers and modelers might find nondeterminism handy; others, though, might prefer a deterministic version.

3. Extra trustworthiness. If we have two versions of the same requirement (one deterministic and the other nondeterministic) we have more assurance that one of them will catch an error. We can also simulate them under the same test suite and see whether they agree on the specification (we will discuss simulating assertions in Chapter 5).

Figure 4.13 shows an example of a deterministic variant of the nondeterministic assertion in Figure 4.8.

Making assertions deterministic can be tricky. It often seems to work for certain scenarios but does not for others. For example, consider the NL requirement:

**R4.2.10:** *Whenever event Q occurs then E may not occur more than N times within the time interval [A,B].*

Figure 4.17a shows a nondeterministic assertion for this requirement. Figure 4.17b is an incorrect deterministic assertion for the same requirement that always records the arrival time of the *latest* Q event. But as evident from the scenario in Figure 4.17c (using N==2 as an example), when the third E is sensed (time t), it is not yet known whether it is better to account it within the [A,B] interval of the first or of the second Q. In this scenario, eventually it is the first Q that has more than N E events within interval [A,B], so the deterministic assertion does not flag a failure when it should.

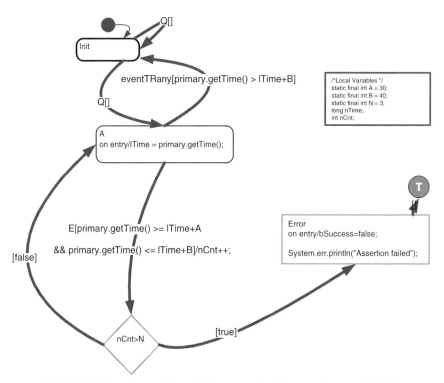

FIGURE 4.17a    A nondeterministic assertion for requirement R4.2.10.

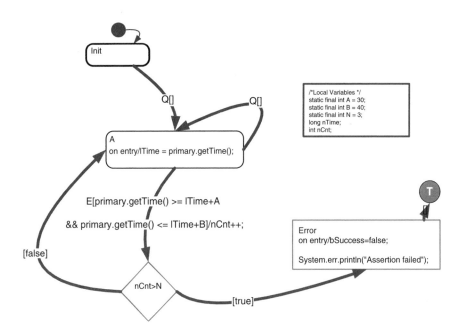

FIGURE 4.17b   An incorrect deterministic assertion for requirement R4.2.10.

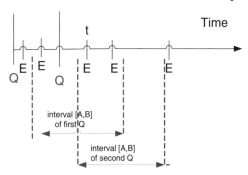

FIGURE 4.17c   A scenario for requirement R4.2.10 (with N==3).

Note, the event named *eventTRany* in Figure 4.17a. It fires while the assertion is in state *A* if the statechart senses any event—e.g., if Q or E fires. Clearly, *eventTRany* induces potential nondeterminism. For example, while the assertion is in state *A*, if event *E* fires, the statechart might be faced with a conflict between the two transitions issuing from state *A*.

## 4.2.11. Debugging Nondeterministic Assertions

As we discussed earlier, a nondeterministic assertion is implemented as an assertion object with a plurality of *Config* inner objects, one per state-configuration object. When animating more than one state-configuration we will see several states and transitions colored or drawn in a certain way. These animated states are often not related; rather the similarity reflects the execution of many *Config* objects. This tends to be rather confusing.

Other StateRover techniques for helping us debug nondeterministic assertions are described below. Consider R4.2 with CNT==3. Figure 4.18a illustrates a nondeterministic assertion for that requirement. Say a JUnit-driven execution induces three state-configurations:

- C0 which is <state==*Init*, nTimer==0, bSuccess==false>,
- C21 which is <state==*Count_1*, nTimer==75, bSuccess==false>, and
- C22 which is <state==*Count_2*, nTimer==25, bSuccess==false>.

Say we are testing a specific scenario and want to monitor *Config* C22 using animation. To do so we first mark a state configuration, in this case C22. We do it from the outside (e.g., a JUnit test case) as in the following JUnit test:

```
assertion.newCar(); // one car
assertion.newCar(); // two cars
String[] sA = new String[]{"Count_2"}; // describe
                                       // states in Configs
                                       // of interest
assertion.markAllComputationsIn(sA); // mark those
                                     // Configs of interest
```

Note how the Configs of interest have been marked at the right time, which is after two *newCar* events have been sensed. Note, too, that once a *Config* is marked all descendent *Configs* are considered marked.

After marking Configs of interest we set animation on only for configurations of interest. In addition, the generated code provides a source code method which is reached only while executing marked Configs. Placing a break-point there lets us debug the generated code knowing which *Config* it is executing for.

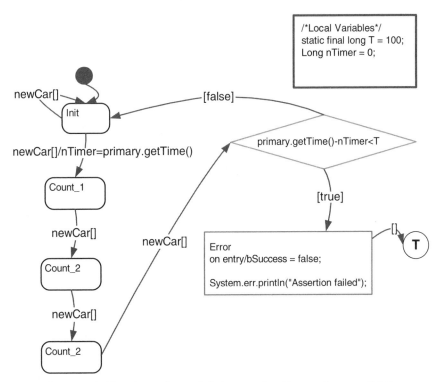

FIGURE 4.18a   A nondeterministic assertion for requirement R4.2.

## 4.2.12. Inverting Nondeterministic Assertions

Consider the following NL requirement, the inverse of R4.2 (whose matching statechart assertion is in Figure 4.8a and 4.18a).

**R4.2.12**: some *T second interval must contain more than CNT new-Car events.*

The statechart assertion of Figure 4.18b implements this requirement. It is almost the same as Figure 4.8a except for the following two changes:

1. The default *bSuccess* logic has been inverted; it is now *false* by default (in the *Init* state) and is set to true in the *Good* state.

2. Priorities are used to invert the default existential behavior of *bSuccess*. The default behavior is that if one *Config* object has *bSuccess* = false then the assertion as a whole projects *bSuccess* = false. Using priorities (state Good has priority 2 and all other states have priority 1) if one state assigns *bSuccess* = true then it must be the Good state, which is always a winner state.

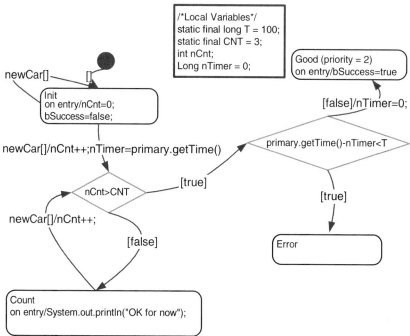

FIGURE 4.18b   Statechart assertion for requirement R4.2.12, the inverse of R4.2.

## 4.2.13. How Do I Write Nondeterministic Assertions?

The technique for writing nondeterministic assertion statecharts consists of two simple steps.

1. Use nondeterminism to "guess" the start of a scenario that violates the requirement.

2. Verify that that is indeed a violating scenario.

For example, in Figure 4.8a: The nondeterministic assertion for requirement R4.2, a violating scenario is one that starts with a *new-Car* event and has more than CNT *newCar* events within a T time interval. Note that this scenario $S$ might be part of a longer scenario. The assertion statechart "guesses" the first *newCar* event of $S$ and then varifies that the remaining part of $S$ violates the requirements. "Guessing" the first *newCar* event of $S$ is done by trying out both possibilities, i.e., that this newCar event <u>is</u> the beginning of a violating scenario **S**, and that it is not.

Likewise, in Figure 4.12: The nondeterministic assertion for requirement R4.2.8a, a violating scenario **S** is one that starts with a *P* event that is accompanies by a *Car* object argument whose reference is *car* and during the subsequent time interval T there is no event **Q** that is accompanied by the same *Car* object. The assertion statechart nondeterministically "guesses" the first *P* event of **S** and then varifies that indeed S violates the requirement. Again "guessing" the first *P* event of **S** is done by trying out both possibilities, i.e., that this *P* event <u>is</u> the beginning of a violating scenario S, and that it is not.

## 4.3. Operations on Assertions

We all know from our programming experience how to compose logical conditions from smaller ones using Boolean connectives, such as *x>0 && y>0* or *x>0 || y>0*. Similar Boolean operations can be performed on LTL formulae.

Statechart assertion compositionality is separated into two cases: primitive assertions with no nesting, and nested assertions. The following discussion specifies statechart assertions for both cases using the notion of assertions as flag raisers discussed in Section 4.1.12— namely, using assertions to indicate failure.

Hence, for primitive assertions, the following holds:

1. *Logical or*, as in NL the following requirement.

   **R4.3.1**: *It is not allowed for either event P or event Q to occur.*

   The assertion statechart of Figure 4.19a realizes this requirement.

2. *Logical and*, as in NL the following requirement.

   **R4.3.2a**: *It is not allowed for event P and event Q to occur simultaneously.*

   This situation is impossible because our model assumes events are never simultaneous.

   **R4.3.2b**: *It is not allowed for both event P and event Q to occur (in any order).*

   The assertion statechart of Figure 4.19b realizes this requirement.

3. *Logical implication*, as in NL the following requirement.

**R4.3.3a**: *If event P occurs then event Q may not occur simultaneously.*

Again, this situation is impossible because our model assumes events are never simultaneous.

**R4.3.3b**: *If event P occurs then event Q may not occur later.*

The assertion statechart of Figure 4.19c realizes this requirement.

For nested assertions, the following holds:

4. *Logical or*, as in NL the following requirement.

**R4.3.4**: *It is not allowed for either requirement P or requirement Q to fail.*

The assertion statechart of Figure 4.19d realizes this requirement using statechart orthogonality, and so does the nondeterministic statechart assertion statechart of Figure 4.19e.

5. *Logical and*, as in NL the following requirement.

**R4.3.5**: *It is not allowed for requirement P and requirement Q to fail.*

The assertion statechart of Figure 4.19f realizes this requirement.

6. *Logical implication*, as in NL the following requirement.

**R4.3.6**: *If requirement P fails then requirement Q must fail.*

The assertion statechart of Figure 4.19g realizes this requirement.

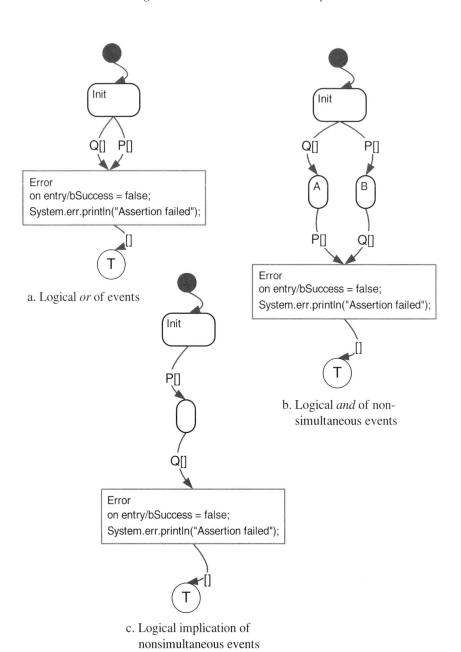

a. Logical *or* of events

b. Logical *and* of non-
simultaneous events

c. Logical implication of
nonsimultaneous events

FIGURES 4.19a–c   Logical operations on primitive assertions.

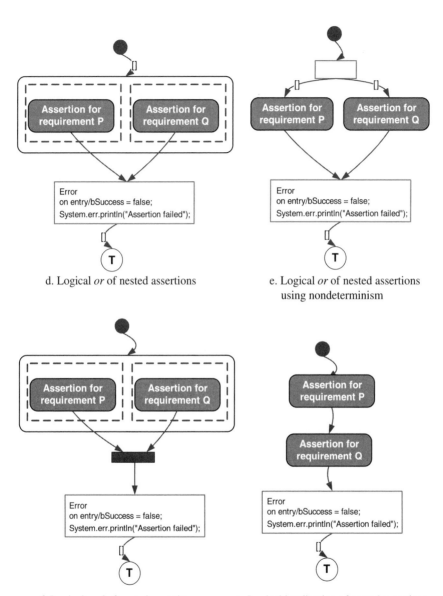

d. Logical *or* of nested assertions

e. Logical *or* of nested assertions using nondeterminism

f. Logical *and* of nested assertions

g. Logical implication of nested assertions

FIGURES 4.19d–g   Logical operations on nested assertions.

## 4.4. Quantified Distributed Assertions

All statechart assertions discussed so far, deterministic and non-deterministic alike, are embedded within a primary statechart and therefore assert about that statechart alone.

In contrast, a distributed assertion can assert about statechart objects other than its primary. Consider a primary *ParkingLot* statechart object, referenced as a *myParkingLot*. Say that *myParkingLot* is associated with a collection of *Car* statechart objects. Consider the following requirement:

**R4.4:** *every Car object in myParkingLot's car collection that is in the Cruising state must enter the Parking state within time T.*

With distributed assertions, illustrated in Figure 4.20, designed to solve these issues, is as follows. First, a new visual construct, a *RemoteAssertion* state, is used, as illustrated inside the *Car* statechart in Fig. 4.20b. Unlike other assertion statecharts, which all have a user-designed statechart specification, a *RemoteAssertion* has no associated statechart. Its job is simple: it holds a reference to the primary object of Figure 4.20a (i.e., to *myParkingLot*) and whenever its own primary statechart object (a *Car* object) enters a state such as *Parking*, it fires *myPrimaryObj.remotePrimaryEntered*("Parking"). The assertion in Figure 4.20a then traverses the transition annotated with that event.

Note there is optionally more than one *Car* object in *myParkingLot*'s car collection. Each has an embedded *RemoteAssertion*, so each will fire *mainPrimaryObj.remotePrimaryEntered*("Cruising") when it enters its *Cruising* state. Say, for example, that the following two *Car* objects fire this method: car #2 and car #7. The *ParkingLot*'s assertion, when triggered by these *remotePrimaryEntered*() events, receives a reference to the *Car* object that invoked the method—namely, either to car #2 or to car #7. This is where nondeterminism kicks in for quantification purposes, just as in Section 4.2.8. The transition loop annotated with the event *eventTRany* spawns two computations, nondeterministically guessing the *Car* object violates the assertion. The computation then visits state *WaitForParking* and stores (in *myCar*) a reference to the *Car* object that entered the *Cruising* state. It then waits for some car to enter its *Parking* state. If a *timeoutFire* fires first, then the assertion is violated. If a car enters its *Parking* state first, then the assertion stores a reference to that car in *aCar*. Then, in the visual switch box, it verifies that *myCar* and *aCar* are identical. If not, then *myCar* is still a candidate car for violating the assertion in the future.

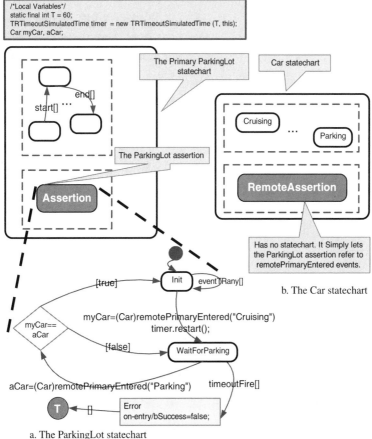

a. The ParkingLot statechart
and its embedded assertion

FIGURES 4.20a–b  A quantified, distributed assertion.

## 4.5. Runtime Recovery for Assertion Violations

As we noted in Section 4.1.12, when an embedded assertion, such as Assertion1 in Figure 4.1a, reaches a terminal state it is typically flagging an error, which is manifested via the *isSuccess()* method. The most common place which acts on this error is an enclosing test suite, as illustrated in Figure 4.2.

Recovery when the primary statechart acts on the flagged error is useful during runtime, if the testing phase has not managed to uncover all violations of the corresponding requirement—namely, a scenario exists that causes the assertion to be violated in runtime. Since we do not have a verification technique that provides 100% test coverage, it is always possible that such a "runaway scenario" exists. Runtime recovery from an assertion violation provides an extra level of protection for such cases.

Runtime recovery is illustrated in Figure 4.1a: when the assertion fails and reaches a terminal state the TLC recovers by moving to *Green* (via the C2 connector).

### 4.5.1. Performance and Reliability Issues

Having nondeterministic assertions execute in deployed, runtime code is a mixed blessing. It can be used for extra trustworthiness, but it also incurs an overhead. Rather than having a single state configuration object, nondeterministic assertions have many. In fact, if we have not sufficiently tested the assertion, we might be unaware that certain scenarios induce an unbounded number of *Config* objects; that is, the longer the scenario, the more *Configs* are generated.

Clearly, then, the efficiency of a runtime nondeterministic assertion is an important issue for many systems. Fortunately, two

StateRover utilities provide some runtime protection against such a situation: the first caps the number of *Configs* during runtime; the second kills marked *Config* objects.

Code generated for deterministic assertions, on the other hand, is as efficient as code generated for the primary statechart, since they use the same code generator. Therefore, it is also as reliable and efficient as the primary code.

## 4.6. The Language Dog-Fight: Statechart Assertions vs. LTL and ERE

The following list summarizes some of the advantages of statechart assertions over LTL and ERE assertions when used with statechart models:

- Statechart assertions are visual, intuitive, and familiar.

- Same formalism used for modeling and specification.

- An event-driven assertion language is used with event-driven modeling language (UML statecharts).

- Stronger descriptive power (Java statecharts are Turing equivalent).

- Better readability. As we will see in Chapter 5 better readability implies better reusability.

- Statechart assertions enable visual and intuitive runtime recovery.

- Statechart assertions enable animation and visual debugging of assertions.

- The source code generated is readily available, readable and for the most part deployable.

- Assertion chaining is simple.

- Natural encapsulation is provided via Java (e.g., a plurality of Traffic-Light Controller objects each with two assertions).

- Generic assertions are readily usable via Java interfaces and event renaming.

- One can assert about states of the primary.

- Simple information passing between primary code and assertion code.

- Built-in support for quantification.

## 4.6.1. Specification Language Conversion

### Specification Language Conversion as a Language War Weapon

Converting an assertion from one formal specification language into another is not recommended, because of various nuances in semantics that are particular to each formalism. Rather, it is advisable to stick to NL and a formal specification language of choice, using the assertion development process, which we will examine in Chapter 5.

Conversion is often the pitfall that people taking the side of any formal specification language fall into. They commonly request that a conversion be performed; for example: *"How would you write this MTL assertion as a statechart assertion?"* Often we would have a hard time doing this because of the different nuances of the particular languages, nuances that are not at all useful towards the *true* end goal of representing real-life requirements. In other words, performing a conversion is a flawed comparison technique. A formal specification language is hardly ever the starting point; NL is. The real question, then, is how easy or difficult it is to represent a NL requirement in one language or another.

We will illustrate below some of the nuances involved in converting LTL or MTL to statechart assertions.

### From LTL to Nondeterministic Assertions

Theory suggests that we can always convert LTL and MTL assertions into nondeterministic assertions using linear blowup. The representation of some unnested LTL and MTL assertions as statechart assertions is rather straightforward, as shown in Figure 4.21.

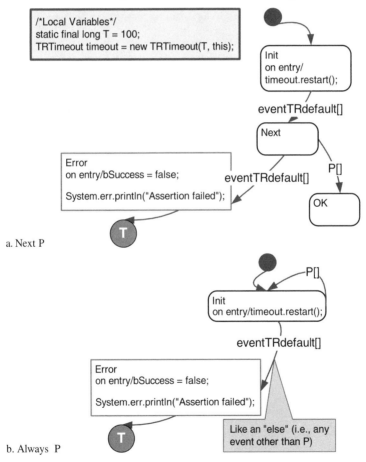

FIGURES 4.21a–b   Statechart assertions for unnested LTL and MTL.

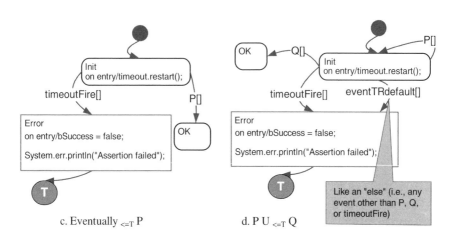

c. Eventually $_{<=T}$ P          d. P U $_{<=T}$ Q

FIGURES 4.21c–d.

The following example illustrates some of the semantic nuances that make specification language conversion confusing. Consider the MTL assertion

**A4.6.1**: *((Always $_{\geq T1}$ P) Implies Eventually $_{\leq T2}$ Q) W R.*

where *P*, *Q*, and *R* are basic propositions. It intuitively covers the NL requirement.

**R4.6.1**: *whenever P repeatedly holds for T1 time units or more then sometime within time T2 time units afterwards Q must hold. This process should continue until R holds.*

*Nuance #1*:  As discussed in Section 3.3.13, *P* in A4.6.1 is a proposition—i.e., a Boolean condition. Statecharts are by and large event based. Therefore, it is likely that because A4.6.1 is forced to use only propositions, it is actually distorting the true nature of the model.

For the sake of discussion, let's face the issue head-on and use a procedural statechart—that is, consider *P* and *Q* as Boolean conditions. Figure 4.22a shows such a statechart that "almost" works: that is, it works except for some language nuances, as discussed below. Note also that the statechart of Figure 4.22a is handicapped; it cannot use events. In general, event-driven statecharts are more readable than precedural ones.

*Nuance #2*: When does the counting of the T2 interval begin? From the *beginning* of the T1 long interval in which *P* holds or from the *end*? MTL semantics for A4.6.1 measure T2 from the starting point of T1 (Figure 4.22b), but is that really the intention of the NL requirement? Probably not; many would argue that the word *afterwards* in R4.6.1 means that counting starts from the *end* of the T1 interval (Figure 4.22c). Stated differently, only when we know that *P* has been true for a sufficiently long interval (T1) we start measuring time for the promised *Q*.

*Nuance #3:* if *R* occurs just after a *T1* period but before the promised *Q*, *Q* must still be true as promised; otherwise A4.6.1 fails, as illustrated in Figure 4.22d. However, the NL requirement wants *R* to behave more like a chop: namely, once *R* is true the assertion has been satisfied. That is achieved using the high-level transition to *Done* in Figure 4.22a.

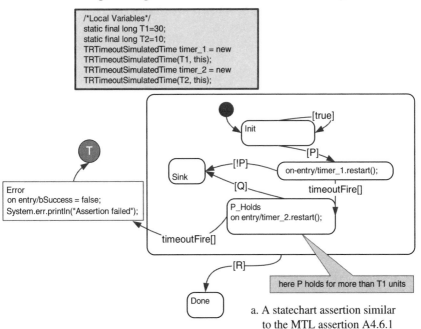

a. A statechart assertion similar to the MTL assertion A4.6.1

FIGURE 4.22a.

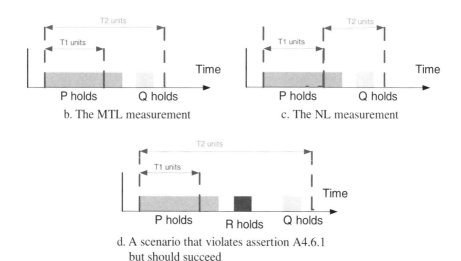

b. The MTL measurement                    c. The NL measurement

d. A scenario that violates assertion A4.6.1
   but should succeed

FIGURES 4.22b–d.

## 4.6.2. From ERE to Nondeterministic Assertions

Figure 4.23 illustrates the straightforward, inherently recursive representation process of EREs as nondeterministic statecharts.

a. ε                                      b. empty set

FIGURES 4.23a–b.

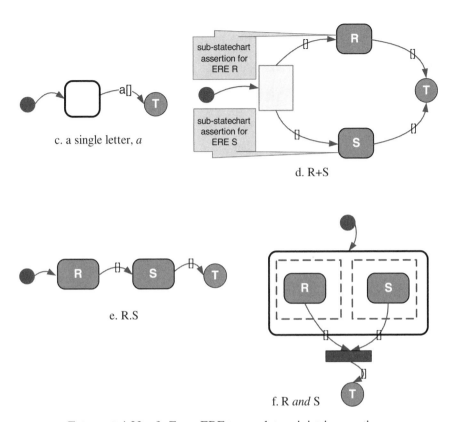

c. a single letter, *a*

d. R+S

e. R.S

f. R *and* S

FIGURES 4.23c–f   From ERE to nondeterministric assertions.

## 4.7. Succinctness of Pure Statechart Assertions

Statechart models and statechart assertions as described in this book are Java statecharts—namely, they use embedded Java variables and code snippets. *Pure statecharts* are not so fortunate; they are mere extensions of FA that include additional features such as hierarchy, orthogonality/concurrence and history.

The cube of Figure 4.24 illustrates known lower-bound and upper-bound results for FA and pure statecharts. For example, deterministic statecharts are, in the worst case, exponentially more succinct than DFAs. In other words, there exists some requirement $R_{det}$ that has a deterministic statechart specification with $n$ states but for which smallest equivalent DFA must have at least $2^n$ states. Similarly, Figure 4.24 shows that nondeterministic statecharts are, in the worst case, exponentially more succinct than NFAs. The edges of Figure 4.24 represent upper bounds as well, such as: every deterministic statechart is convertible to an equivalent, though exponentially larger, DFA.

Note that these lower bounds should be used with a grain of salt because they are existential in nature and not necessarily generalizable. For example, a devil's advocate would claim that there are only a few requirements like $R_{det}$ while the vast majority of requirements are not so demanding. Also, he would argue that $R_{det}$ and its kind are uninteresting requirements that do not pertain to the real world; for example, a proof for NFA-related lower bounds that uses an exponentially large alphabet. Furthermore, a devil's advocate would claim that restricting ourselves to pure statecharts rather than using Java-statecharts is the culprit.

Nevertheless, these bounds do give a hint at the core capabilities of each formalism. In fact, in Section 4.2.9, we described a practical requirement (4.2 and its variant R4.2.9.1) and then we compared the two approaches, namely, the deterministic Java-statechart assertion (Figure 4.13) and its nondeterministic statechart equivalent (Figure 4.8). The nondeterministic statechart assertion version is arguably more readable than its deterministic counterpart for reasons rooted in the current succinctness discussion.

The primary statechart feature that contributes to these lower bounds is orthogonality/concurrence. In other words, these results show that statechart *orthogonality/concurrence* acts as a third dimension, independent of existential and universal nondeterminism.

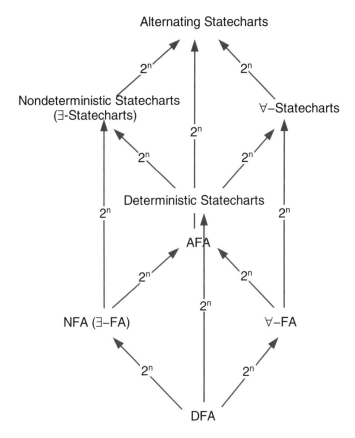

FIGURE 4.24   The lower bound (succinctness) cube for FA and pure statecharts.

## 4.8. Temporal Assertions vs. JML and Java Assertions

JML is a Java-related interface specification language. The following listing shows a JML annotation for the Java method mySqrt(). It describes a precondition (*requires y >= 0*) and a postcondition (*ensures…*). In other words, the JML specification describes a contract for the mySqrt() method:

```
/*@ public normal_behavior

@ requires y >= 0;

@ ensures \result * \result <= y

@ && y < (\result + 1) * (\result + 1);

@*/

public static int mySqrt(int y)

{ ...

}
```

The JML contract for mySqrt() is for every particular call made to the method. Using LTL we would have written it as a pair of LTL assertions:

1. *Always* $(y > 0)$

2. *Always*( \result $\leq y$ && $y < (\result + 1) * (\result + 1))$

A Java assertion is a Boolean logic statement written within a body of Java code, plus an associated action, such as:

*assert x < 0 : "x is not < 0";*

At this point, we should note two common properties:

1. Both JML and Java assertions use Boolean propositional logic for specification.

2. Both techniques have good reasons for isolating their assertions from the main body of code. One primary reason is that we need to be able to turn the assertions on and off, depending on the context of execution (e.g., verification or testing time vs. deployed code).

As for the first point, note that preconditions, postconditions and invariants are merely propositional logic properties that must be true every time a particular location in the code is reached. As such they are equivalent to LTL assertions that use one operator, *Always;* have no nesting; and have no real-time constraints.

Temporal assertions, written as statechart assertions or variants of LTL, are also meant to be used in a isolatable way—i.e., they can be turned on and off at will. However, temporal assertions have the ability to capture requirements that go beyond pure propositional logic. For example, consider the NL requirement:

**R4.8**: *Whenever y < 0 then within one minute: (a) \result > 100 and (b) thereafter, for 5 minutes, \result > 50.*

It captures the relationship between the input and the output (*y,* and *\result* ) *over time.* This assertion is expressed in MTL as:

**A4.8**: *Always (y < 0 Implies Eventually $_{< 1 \, minute}$ (\result > 100 And Always $_{>= 5}$ \result > 50))*

To achieve the same effect using a JML or Java assertion, we would essentially have to build a machine that implements this or an equivalent statechart assertion. It can be done, but it is highly unproductive because it's a time-consuming and error-prone process. In fact, the assertion would then be as likely to have errors as the body of code it's supposed to be protecting.

## 4.9. Commonly Used Assertions

This book contains statechart assertions for many common NL requirements. Some examples are:

- Figures 4.3a and 4.3b contain statechart assertions for the NL requirement R4.1.6: A *P event is not permitted within the T time interval following an event Q.*

- Figure 4.3d contains a statechart assertion for the NL requirement R4.1.6b: *A P event is not permitted after T time of an event Q.*

- Figure 4.3f contains a statechart assertion for the NL requirement R4.1.6.1: *A P event must occur within T time of an event Q.*

- Figure 4.5 contains a statechart assertion for the NL requirement R4.1.9: *No more than N newCar events are allowed within the T time interval following an event Q.*

- Figure 4.8a contains a statechart assertion for the NL requirement R4.2: *No more than N (N > 0) events E can be sensed within a single T second interval.*

- Figure 4.15a contains a statechart assertion for the NL requirement R4.2.9.3a: *No more than N (N > 0) events E between event Q and event R.*

- Figure 4.16 contains a statechart assertion for the NL requirement R.4.2.9.4: *Whenever event Q occurs then P is not allowed outside time interval [A,B].*

- Figure 4.19 contains statechart assertions that emulate basic LTL and MTL assertions.

In addition, Figure 4.22 contains a few more common statechart assertions for common NL requirements:

- Figure 4.22a for: *Whenever event P occurs then event Q must occur within time T.*

- Figure 4.22b for: *Whenever event P occurs then event R cannot occur before event Q.*

- Figure 4.22c for: *Whenever event P occurs then event R must occur before event Q.*

- Figure 4.22d for: *Whenever event P then condition C must hold until event R.*

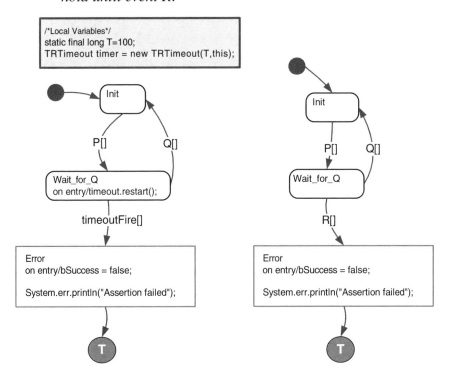

a. Whenever event P occurs, then
   event Q must occur within time T.

b. Whenever event P occurs, then
   event R cannot occur before event Q.

FIGURES 4.22a–b.

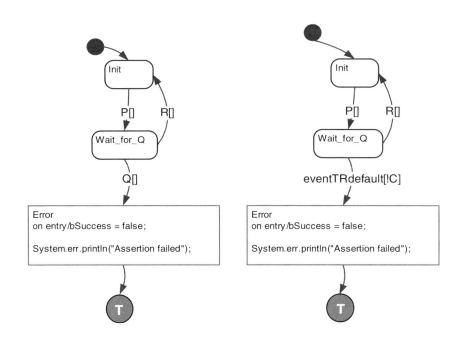

c. Whenever event P occurs then event R
    must occur before event Q.

d. Whenever event P then condition C
    must hold until event R.

FIGURES 4.22c–d.

# Chapter 5

## Creating and Using Temporal Statechart Assertions

With classical verification formal methods, such as model checking and theorem proving, as well as with many runtime monitoring tools, the distinction between a formal specification of a reactive system and the system itself is clear: formal specification is written in some unfamiliar logic (e.g., LTL or branching time temporal logic), whereas the system is described as a state machine of sorts or a C, C++, or Java program. However, this book advocates using statechart assertions for specification, which begs the question: Why not have both aspects of the system, statechart model and formal specifications, reside in the model? In other words, why do we need statechart assertions at all? Can't the assertions live as part of the model? Aren't they duplicating what is already described in the model?

### 5.1. Motivation, or Why Use Temporal Assertions?

There are numerous reasons for keeping assertions separate from the model, regardless of the language used to capture them. The following subsections contain a few motivations for creating assertions and for separating them from the model.

## 5.1.1. Look at Them: They're Doing the Same Thing

As we discussed in Section 4.8, JML and Java assertions are widely used to verify contracts. In those cases we could ask the same kind of question, namely: *Why isn't a Java assertion written as plain Java just like any other part of the program?* In other words, *Why is the assertion even needed?* Here are some answers:

1. We need to be able to turn the assertion on and off depending on the context of execution, e.g., testing time or deployed code.

2. We need a language-based utility to identify the contract specification. The utility helps ensure that even if the implementation changes later, the contract still remains. Stated differently, that the baby (assertions) is not thrown out with the bath water (implementation code).

3. There are three types of temporal assertions—test-time, runtime, and simulation-time—which we will discuss below. Often, the second and third cannot be integrated with the main body of code: (i) runtime assertions might actually live in a separate body of code called the *safety executive*, and (ii) simulation-time assertions are never active in the deployed code.

## 5.1.2. Verifying Temporal Contracts

Assuming that we agree with the need for JML assertions or Java assertions, temporal assertions provide similar functionality, but for properties that are more complex, namely those that require order and sequencing, as we discussed in Section 4.8.

## 5.1.3. Assertions as Automated Monitors

The conventional test process for reactive components comprises two primary testing activities: test generation and test monitoring. Test generation for a reactive component amounts to the generation of sequences of input events, conditions, timing information, and data objects to the statechart (the statechart under test, or SUT). When the test sequence is applied to the SUT, a human observer observes the sequences of outputs generated by the statechart controller and decides whether the behavior is good or bad. Each such test is called a *test-case*.

With the advent of tools such as JUnit, the results of this manual testing process are recorded for later use using extensive test-suite collections of test-cases. Subsequent invocations of the test-suite are then performed automatically using this existing test-suite.

The automated approach to testing resembles the manual approach, as shown in Figure 5.1, except that instead of using a human monitor we use an automatic monitor based on temporal assertion monitoring. An additional improvement is achieved when automatic test generation is used in addition to manually constructed test sequences. (We will discuss this topic in detail in Section 5.7.) With the volume of tests generated by an automatic test generator, anything but automatic monitoring is impractical.

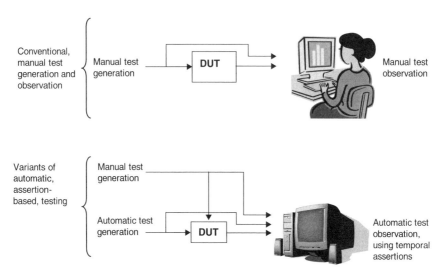

FIGURE 5.1   Automatic testing vs. manual testing.

## 5.1.4. Assertions as Champion Representatives of Individual Concerns

Legislation in the American political system is the steady-state result of many political forces, each vying for the interest of a specific concern or cause. Lobbyists are often the political champions of a cause. Consider a particular bill, such as the highway bill of 2005: a complex piece of legislation stretching over more than a hundred pages. Now think of Bob, a lobbyist working for a well-known automaker. Bob's assignment is to represent the company's desire to see that single-occupant hybrid vehicles are allowed to use carpool lanes. Hence, within the diverse universe of forces influencing the bill, pushing and pulling in various directions, Bob is representing a *particular* interest, or concern.

Similarly, albeit with more dignity, an assertion is a representative of a single concern within a complex system. It makes sure that the concern is not forgotten, lost, or misinterpreted over time. The system is the implementation that results after all concerns that have a say have been considered.

Consider as an example the traffic light controller of Figure 5.2 that controls a main road and a secondary road and issues output color commands as pairs, such as <Red, Green> for red in the main direction and green in the secondary direction. A TLC *cycle* is defined as a change from red to green and back to red. Three concerns for this controller are, in NL:

1. NEVER_GREEN_TOO_SOON: *light color in main direction never changes from red to green unless it was red for at least 30 seconds.*

2. ALWAYS_EVENTUALLY_CHANGE: *whenever light color in a particular direction is red it should eventually, within at most 5 minutes, turn green.*

3. AVOID_FLICKER: *the lights in the main direction should not be green for less than 10 seconds in more than 3 consecutive cycles.*

An input contract assertion named *NewCar_Contract* verifies that *newCar* input events are spaced at least 10 seconds apart. In addition, the TLC counts cars (three cars waiting while the lights are red in one direction causes them to change to green) and also operates a camera. The Appendix contains the TLC statechart with diagrams and code for these three assertions, along with JUnit simulation scenarios.

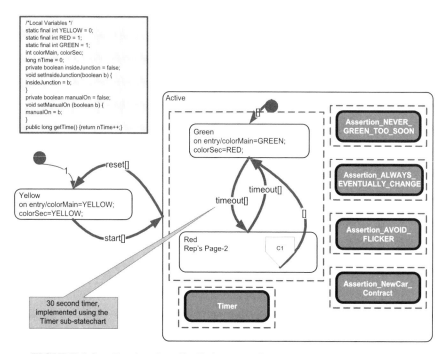

FIGURE 5.2a  Top level traffic light controller with embedded assertions.

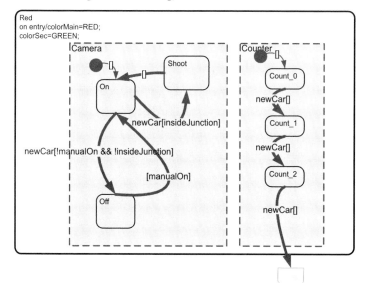

FIGURE 5.2b  Contents of the *Red* state.

The TLC statechart tries to accommodate all requirements, those specified as assertions and those that are not. For example, it chooses a one-minute timeout interval for toggling between the *Green* and *Red* states, although no requirement says so explicitly; it is a designer's choice within the allowable framework governed by the requirements. The designer realizes the NEVER_GREEN_ TOO_SOON requirement relying on the 10-second interval between cars promised by *NewCar_Contract,* having calculated that three cars ten seconds apart amounts to thirty seconds.

Note how the TLC per se has no explicit representation of the NEVER_GREEN_TOO_SOON requirement. As a result, a simple change to the TLC could inadvertently induce a violation of this requirement; for example, changing the input constraint allowable between *NewCar* events from ten to five seconds. The NEVER_ GREEN_TOO_SOON assertion, on the other hand, is an explicit representation of this cause. In fact, the TLC has a bug, as discovered by the TestTLC JUnit test case in the appendix. The bug is that the input contract guarantees ten seconds only between *NewCar* events but not before the first one, so the sum total delay produced by three consecutive *NewCar* events might be less than thirty seconds.

From the theoretical standpoint of Chapter 1, the TLC statechart has three input alphabets: an event alphabet that contains *timeout*, *reset*, and *start* events, an alphabet with insideJunction and its inverse, and an alphabet with manualOn and its inverse. The TLC's output alphabet consists of the output color commands mentioned earlier.

As we saw in Chapter 1, the universe for the TLC statechart, as well as for the assertions, is a set that consists of all of possible input-sequence, output-sequence pairs. The TLC, however, responds to an input sequence $S_{in}$ with one output sequence $S_{out}$, and no more;

we say that the TLC *generates* the pair $S_{in}$, $S_{out}$. Clearly, not every input-sequence, output-sequence pair is generated by the TLC. For example, any pair of sequences in which a light is red continuously for 6 minutes in one direction is not generated by the TLC. We do, however, expect the TLC to generate some output sequence for every input sequence. In other words, the universe provides for the possibility that any input sequence will occur. Certainly, an input contract might preclude some input sequences ever reaching the TLC statechart, but it is then the responsibility of an input-assertion to verify that the contract is abided by.

The statechart model and the assertions behave in slightly different ways:

- The TLC-statechart model is expected to generate only allowable output-sequences.

- Assertions are expected to flag bad behavior—i.e., to set *isSuccess*()=false for any unallowable input-sequence, output-sequence pair.

The Venn diagram in Figure 5.3 shows the relationship between the universe (the set Σ* of all possible sequence pairs), the TLC statechart, and the assertions. The TLC statechart maps one output sequence to a given input sequence; it is therefore a subset of the universe. Similarly, every assertion covers a subset of the universe: its bubble shows the input-sequence, output-sequence pairs the assertion allows. Unallowable pairs would therefore be captured by the complement set, namely the area outside the rectangle.

Note that rectangles denote assertions whereas the bubble for the TLC is drawn somewhat like an amoeba, being a heterogeneous sum of all concerns. The rectangles for assertions always include the TLC-statechart amoeba because they are *weaker* than the statechart—that is, more general. In other words, a logical implication goes in the following direction: if the statechart produces some input-sequence, output-sequence pair, the assertion should evaluate to *true* for that pair (but the inverse doesn't necessarily hold).

Stated differently, the inverse of an assertion (captured visually by the areas *outside* its rectangle in the Venn diagram) is always *stronger* than the statechart; i.e., if an assertion evaluates to false for some <input-sequence, output-sequence> pair, the TLC-statechart should not generate that pair. Assertions are usually thought of in this sense, called the *negative interpretation*; that is, assertions are acted on when they fail.

Rectangles for assertions can overlap. For example, in Figure 5.3, the assertion ALWAYS_EVENTUALLY_CHANGE overlaps the bubble for the assertion AVOID_FLICKER. However, not every behavior that satisfies AVOID_FLICKER works for ALWAYS_ EVENTUALLY_CHANGE. As engineers, we are interested in capturing input-sequence, output-sequence pairs that satisfy *all* assertions; we're interested in the intersection of the squares. Hopefully, the entire TLC's bubble is fully included within the resulting intersection, and it is precisely the job of runtime monitoring to verify that, as we will discuss in Section 5.4.

This visual representation shows that each assertion covers a specific, typically coherent concern. Sometimes there is some overlap between concerns; that's all right. But all the concerns should be realized by the TLC statechart, which is in effect *the sum total of all concerns*.

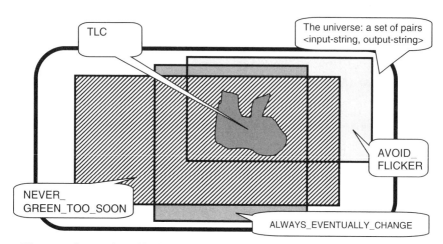

When assertions are in positive form, namely, if an output-string out is generated by the model in response to an input string in than the assertion evaluates to true given the pair <*in,out*>.

FIGURE 5.3   The relationship between the universe,
the TLC statechart, and the assertions.

## 5.1.5. Negative Information

Consider the TLC's NEVER_GREEN_TOO_SOON requirement. We could implement it in a variety of ways. For example, we could have two consecutive 15-second delays because of a separate concern that requires a measurement (of the number of cars waiting in the direction of the red light) be performed after 15 seconds. In this case, the 30-second delay *is* implemented, even though there is no explicit statement in the body of code of what the *actual* requirement is. It is in fact a requirement about something that should *not* happen, called a *negative information* requirement; the statechart controller is (hopefully) implementing its positive counterpart. Computer scientists would probably say that the negative information requirement could easily be converted into a positive form. That is true when the requirement is written in a formal language or logic, but, as we will see in Section 5.3, requirements are often originally conceived as NL requirements. So what would the positive form of our safety requirement be? Would it be: *whenever light is Red for less than 30 seconds, then it should remain Red for one more second*? This wording amounts to using recursion in NL. Even if we have a more reasonable positive form for the requirement, it's your cognitive solution, not mine. My way of thinking led me to feel comfortable with the negative information style requirement, and therefore it is that form that I would represent as an assertion.

## 5.1.6. Assertions in the Context of Statechart Model Changes

Imagine a coffee maker and the NL requirement:

**R**$_{coffee}$: *do not accept money for a future order while making a cup of coffee.*

The developer of the coffee makers' statechart controller claims that $R_{coffee}$ is conformed to in the most obvious way: in the state where the statechart starts making coffee, she assigns *acceptMoney = false*, with *acceptMoney* being the variable controlling money acceptance.

The developer is making a valid argument as long as she is working alone on a project that is not evolving, such as a student project. Real-life projects are different though. They tend to evolve from version to version, engineering concerns change over time, and multiple users will modify the statechart either simultaneously or in the future. With many different modifications to the statechart model taking place over time, someone could overlook the requirement $R_{coffee}$ and make a modification that violates it.

### 5.1.7. Assertion Reuse

Consider the NL requirement:

**R5.1.7**: *whenever there is a request, then an acknowledgment should be received within time T afterward.*

This is a fairly generic temporal requirement we might find in many projects and often within a project. Having assertions set aside for such requirements facilitates reuse. Indeed, many assertions are rather generic and therefore have the potential of being reused.

Two examples of libraries of reusable assertions are the Kansas State University specification patterns library, which uses textual representations, such as LTL, and the StateRover's library of statechart assertions.

### 5.1.8. Nondeterministic Assertions

As we discussed in Section 4.2, some assertions are conveniently described as nondeterministic statecharts. Obviously, with sufficient time and money we could launch a project for each such assertion and make it deterministic. Nevertheless, nondeterminism provides a mechanism for creating simple and succinct assertions for rather complex requirements.

Nondeterminism, however, is useful only when used within an assertion, and cannot be used within a statechart model.

### 5.1.9. Early Requirement Simulation

One of the most appealing features of formal specification is the ability to simulate requirements during early phases of the design process. (We will discuss this topic in Section 5.3.4.)

## 5.1.10. Requirement Tracking

Assertions are useful for requirement tracking: since each assertion represents a requirement, assertion monitoring, during testing or runtime, amounts to requirement tracking. And since assertions are coherent and separate from the conglomerate, heterogeneous statechart model, requirement tracking can be used to identify the precise part of the program responsible for verifying a given requirement.

## 5.2. Applying Assertions: Three Uses

Assertions can be applied in three ways:

1. **Testing time assertions**. This is the conventional way Java assertions, JML contracts, and REM are used; namely, assertions exist, or are turned on, only during testing.

2. **Simulation time assertions**. Here, assertions are used only when the statechart model is in a simulated environment. In this environment, we can simulate otherwise impossible scenarios with specific assertions written for those scenarios alone. For example, consider the Ballistic Missile Defense project. In this project it is natural to require that *no enemy missile will ever hit the White House*. Clearly, unless we simulate the necessary scenarios we will probably, and hopefully, never get to examine such an assertion, and if we do it will be too late. In a simulation environment, however, we can simulate the launch of an enemy missile attack on the White House and see whether or not the system fails the assertion.

3. **Runtime assertions**. Runtime assertions are deployed in the release code and executed on the end-user's machine. A runtime assertion provides an additional layer of safety over the model statechart by triggering custom recovery whenever it fails. In Section 5.5 we describe two techniques for runtime recovery from violations of runtime requirements.

## 5.3. Writing Assertions

### 5.3.1. Positioning: Writing Assertions from a Tester's Perspective

A primary concern of every software tester is to create tests that add value. One pitfall that is easy to fall into is to replicate what the SUT is already doing. Consider the statechart in Figure 5.4a, for example. Now, consider the NL requirement *whenever the state-chart (SUT) is in state A and receives event req, then eventually—within 10 seconds—it moves to state B.* While the requirement makes sense, having an assertion for it (Figure 4.5b) does not. This assertion does not add any value beyond what the model already specifies.

The issue with such an assertion is that it is a white-box assertion; namely, it checks how the SUT is going about its job, rather than checking whether it's doing the right job to begin with. The assertion asserts about the inner structure of the model statechart SUT (the transition between A and B) rather than interesting artifacts visible from outside the box. In other words, assertions should be constructed from a black box perspective, the same way a human tester is expected to write tests for the SUT.

In terms of the example in Figure 5.4, we might argue that the assertion above makes sense because the *ack* message is generated in state B. In that case, however, the following NL requirement, which we will denote R1 is a more accurate requirement: *whenever event req is sensed, then eventually-within-10-seconds ack is generated.*

Nevertheless, assertions might exist only in the *context* of specific SUT states. For example, what if R1 is too general because we actually mean R2: *whenever event req is sensed while in state A, then eventually-within-10-seconds ack is generated*? This is precisely where the context of the assertion can be used, as illustrated in Figure 5.4a and 5.4b. Here Assertion1 for R1 is used only in the desired context, namely only after state A is entered. Thus, the assertion is asserting about the black box interface (*req* and *ack* events), but it is restricted to a specific context of the SUT.

Figure 5.4 raises the obvious question: since we've restricted the assertion to the context of state A, how do we verify that the SUT actually enters state A "correctly", that is, precisely under the desired circumstances? There are three possible answers:

1. We trust that that aspect of the SUT—namely that it enters state A "correctly"—is verified using conventional methods and doesn't require an assertion.

2. We could write the actual requirement (or requirements) for entering state A "correctly," using the alphabet of the black box interface, such as: *whenever event eBegin is sensed, then state A should be reached within 1 second.*

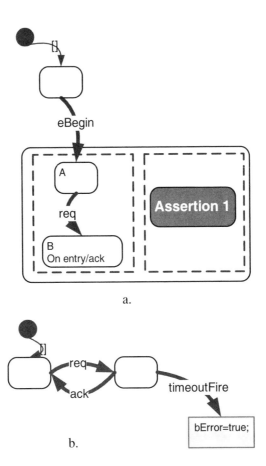

a.

b.

FIGURES 5.4a–b   A statechart with an embedded assertion statechart.

Another aspect of white-box versus black box assertion writing has to do with automatic test generation. In Section 5.7.3 we will consider an automatic white-box test generator that can pump a huge number of tests into a statechart SUT. The tests, however, are useful only if an automatic observer—namely assertions—verifies the outcome of each test against some expected behavior. But if an assertion is merely a replication of part of the statechart SUT, it will never fail, rendering the entire automatic testing approach useless.

## 5.3.2. Assertion Types

OK, we know that assertions should *not* replicate the SUT statechart's logic, but what *should* they be? Remember, again, that for the most part, an assertion is the representation of a single, coherent concern. A primary concern of any component is its interface, or input-output content. Temporal assertions are used to specify contractual time and sequencing constraints.

There are three types of *temporal contract assertions*:

1. Temporal contracts concerning valid input sequences. For example, in the TLC in Section 5.1.4 successive *newCar* events are guaranteed to be at least ten seconds apart. These assertions are called *input assertions*.

2. Temporal contracts concerning valid output sequences, namely contracts about sequences of outputs the statechart generates. A traffic-light controller asserting that the red light is never on too soon after the green light, light colors being outputs, is an example.

3. Temporal contracts concerning valid input-to-output sequences, namely contracts about what the statechart controller should do in response to given input sequences. Assertion A5.3.3 is an example.

Clearly, the first two types are special cases of the third. In addition, there are two other types of assertions:

• Negative information assertions (which we discussed in Chapter 3). These assertions explicitly express a specific concern regarding a certain behavior that the statechart controller should *not* exhibit. The statechart controller, on the other hand, tries its best to comply with the requirement, but never *explicitly* captures it because it has no way of doing so. Consequently, the negative information requirement needs an assertion to champion it.

- Statistical assertions. These assertions are expected to fail from time to time and exist to collect such statistical information as the frequency of assertion failures, or the average time between failures. As an example, consider the following NL *flicker* assertion for the traffic light controller:

**R5.3.2**: *lights in the Main direction should not be green for less than 10 seconds.*

We can expect this assertion to fail during testing and runtime. Nevertheless, statistics about the failure of this assertion are useful. This is especially true if a very large sample set of tests is used, as the case for automatically generated tests discussed later in this chapter.

Note that statistical assertions are not related to customer requirements. Therefore, they do not participate in a regression test suite.

Temporal assertions are also classified according to their use time, as discussed in Section 5.2.

### 5.3.3. From the Human Mind to Correct Assertion

Figure 5.5 shows the development process for assertions. It is best that the entire process be performed by the same individual (or team), which we will call the *assertion developers*. The assertion developers should be in close contact, or include, domain experts.

The first step for the assertion developers is to come up with a conception of the requirement. This step is often done during the development of a UML use-case, its main success scenario, or secondary scenarios.

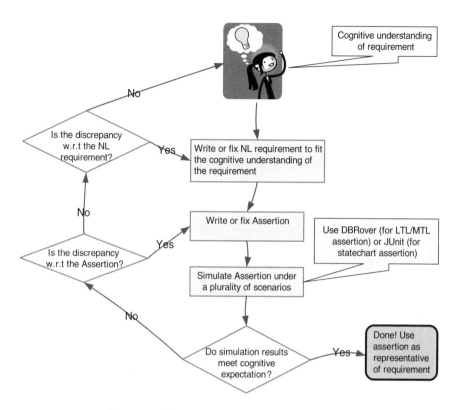

FIGURE 5.5  Assertion development process.

In the second step, the cognitive requirement is expressed in NL for three purposes: (i) to capture it, since in most organizations if it's not written as NL, it doesn't really exist as a requirement; (ii) for documentation; and (iii) for peer and management review.

In the third step, the assertion developer writes the NL requirement as an assertion, such as a statechart assertion or a temporal logic assertion. A devil's advocate might argue that if we create an assertion, the second step is redundant. Perhaps that is true logically, but it is not very practical: NL documentation and specification are established and required components within the development of complex systems, a fact that is not going to change for a long time.

The fourth step is simulation, which we'll discuss in greater detail in the next section. For every scenario being simulated the assertion developer compares the simulation results with the behavior she cognitively expects, based on her understanding of the requirement. If she is not satisfied with the results there are three possibilities:

1. The assertion is an incorrect representation of the NL requirement.

2. The NL requirement is an incorrect or ambiguous representation of the cognitive requirement.

3. The cognitive requirement was not thought through well—namely, she didn't think of this specific scenario when she developed it.

Let's illustrate the process with an example. Obviously, this is not the place to consider the cognitive process so we will begin with the NL requirement, which is:

**R5.3.3:** *Request and ack are Boolean variables. When the request becomes true, then within one minute an ack should be generated and should be continuously true until request turns false.*

The following MTL assertion tries to capture this requirement:

**A5.3.3:** *Always (request implies Eventually $_{\leq 1\text{-minute}}$ (ack Until Not request))*

The assertion statechart in Figure 5.6 likewise tries to capture the requirement. As we shall see the MTL and the statechart assertion are similar but not behaviorally identical.

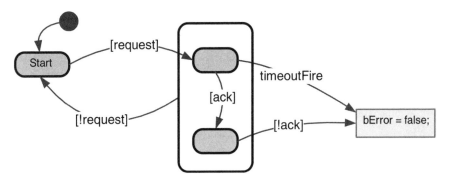

FIGURE 5.6   The statechart assertion for requirement 5.3.3.

Now consider a simulation for scenario #1 illustrated by the timing diagram in Figure 5.7a. The scenario will induce a failure in the statechart assertion and a success in the MTL assertion. The reason is that the MTL assertion allows "false alarms" in the sense that it doesn't mind seeing two candidate *ack*'s, the first one becoming false too soon, but the second one satisfying the requirement. The statechart assertion on the other hand expects the first candidate *ack* to satisfy the requirement.

Which of the two assertions is correct? The NL requirement does not address this issue, it does not require a unique *ack* nor does it specifically allow "false alarms." It's also possible that the assertion developer simply overlooked the issue of whether *ack* should be unique, perhaps because she never considered this scenario. The point is that when the behavior exhibited in a simulated scenario doesn't match the assertion developers' expectation it is not necessarily because of a bug in the assertion. The cause might be an error or ambiguity in the preceding informal steps including incomplete, contradictory, or ambiguous cognitive specification.

Next, consider scenario #2 in Figure 5.7b. Scenario #2, too, induces a success for the MTL assertion, because *Not request* is true within one minute of *request*, thereby satisfying *ack Until Not request* within one minute of *request*. Similarly, scenario #2 induces a success for the statechart assertion. However, do the two assertions really represent the assertion developer's cognitive understanding of the requirement under this scenario? Perhaps she believes that even if request subsides sooner than within a minute, *ack* must nevertheless be generated within a minute. Perhaps she simply overlooked this extreme case.

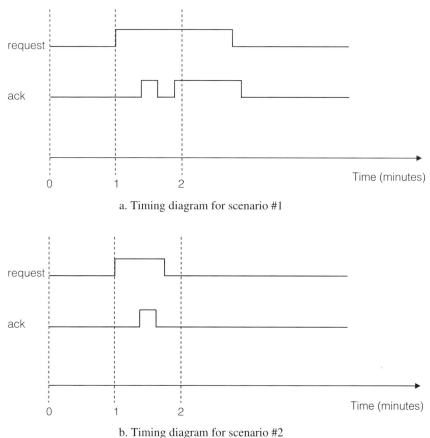

a. Timing diagram for scenario #1

b. Timing diagram for scenario #2

FIGURES 5.7a–b   Timing diagrams for R5.3.3.

## 5.3.4. Simulating Assertions

We considered the DBRover simulator for LTL and its dialects in Chapter 3. Figure 3.6 illustrates the DBRover simulation of assertion A5.3.3. The user begins a scenario by setting basic propositions to true (1) or false (0) along a timeline. That yields a representation similar to a timing diagram. The DBRover scenario also contains real-time constrains and time-series constraints, as shown in Figure 3.6.

Statechart assertions are simulated using the generated code and the JUnit testing framework. The Java code below, for example, consists of a JUnit test (simulation) of the assertion in Figure 4.1b under a specific scenario:

```
import junit.framework.*;

public class TestAssertion extends TestCase {

    private Assertion assertion = null;

    protected void setUp() throws Exception {

        super.setUp();

        /**@todo verify the constructors*/

        assertion = new Assertion();

    }


    protected void tearDown() throws Exception {

        assertion = null;

        super.tearDown();

    }
```

```
public void testExecTReventDiapatcher() {
    assertion.newTruck();
    assertion.newCar();
    assertion.newCar ();
    assertion.newCar ();
    assertion.newCar ();
    assertFalse(assertion.isSuccess());
        // assertion is expected to fail
        //because more than N newCar's were
        //detected after the newTruck
    }
}
```

A JUnit test suite for an assertion therefore serves as the repository of scenarios used to evaluate the assertion from as many angles as possible. It also serves as a basis for the SUT's test suite.

### 5.3.5. Assertion Libraries

One of the reasons we discussed for having assertions separate from the statechart controller they are monitoring is assertion reuse, and therefore the value of assertion libraries, either for a single project, or perhaps universal.

One of the two libraries that we mentioned is the Kansas State University (KSU) specification patterns library, known formally as the *Specification Patterns for Finite-state Verification Library*. It contains patterns written in several languages, including LTL, that are clearly generic and contain no real-time constraints, for example.

The following is an example from the KSU library as provided in NL and LTL:

**R5.3.5**:

```
transitions to P-states occur at most 2 times before R.
```

**A5.3.5**:

```
Eventually R Implies ((! P && ! R) U (R | ((P && !R) U

(R || (((!P && !R) U (R || ((P & !R) U (R || (!P U

R)))))))))
```

The library has two serious drawbacks. The first is the absence of a simulation test suite for the assertions. We can seldom be sure we truly understand a requirement sufficiently to just use it verbatim, not to mention to modify it, as illustrated by assertion A5.3.5, which is evidently hard to read and to reuse. To understand its behavior we would want to play with it, or exercise it with pre-existing scenarios. The second drawback is the lack of constraints. Adding real-time constraints, such as constraining the eventuality in the LTL formula *Always p Implies Eventually q,* seems trivial. But in most cases we would not make a change—any change—to a program and not test the program afterward. Similarly, after adding real-time constraints to an assertion, we would like to test it, and we need a test suite to do that.

The other library that we mentioned, the StateRover, is a small, extendible library of both deterministic and nondeterministic state-chart assertions. Each assertion is accompanied by a JUnit test suite. Figure 5.8 shows a library assertion whose NL specification is: *no fewer than N events E between event Q and the first event R thereafter.* The following is the library's JUnit test for this assertion, using N==3.

```java
import junit.framework.*;
public class TestAssertion1 extends TestCase {
    private Assertion assertion = null;
    protected void setUp() throws Exception {
        super.setUp();
        assertion = new Assertion();
    }

    protected void tearDown() throws Exception {
        assertion = null;
        super.tearDown();
    }

    public void testExecTReventDiapatcher() {
        assertion.Q();
        assertion.E();
        assertion.E();
        assertion.E();
        assertion.R();
        assertTrue(assertion.isSuccess());
    }
}
```

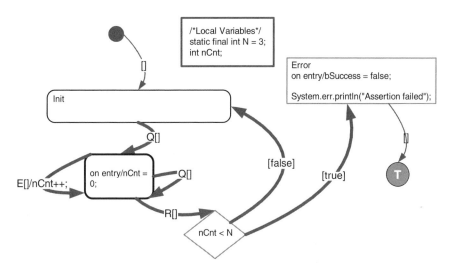

FIGURE 5.8   Statechart assertion for NL specification: *no fewer than N events E between event Q and the first event R thereafter.*

## 5.4. Runtime Execution Monitoring— Runtime Verification

Recall our Runtime Execution Monitoring (REM) discussion from Chapter 3. To repeat, REM executes specification assertions and notifies the user or the enclosing test suite manager of the success or failure of the specification in that particular execution, as shown in Figure 3.2. REM of statechart assertions is no more than the execution of the code generated from a statechart model that contains embedded statechart assertions, like the statechart in Figure 4.1b.

REM is an essential component of test automation. As illustrated in Figure 5.1, REM can be used to monitor the response of a program to manually written tests and to automatically generated tests. REM can also be used in runtime to monitor and possibly recover from violations to runtime assertions.

Temporal Logic REM tools come in two primary flavors: local and remote. Both observe basic propositions in the program being monitored (the SUT), typically via some sort of instrumentation.

In Section 3.3.12, we discussed the Temporal Rover local monitor. Again, it generates source code (C, C++, or Java) for an assertion and embeds it inside the SUT's code as a function or a method. When the combined code executes Boolean values for the basic propositions and the values are passed into the assertion's block of code, the assertion is evaluated, a custom action is optionally executed, and control returns to the SUT.

Remote temporal logic monitors the SUT with code that samples the values of basic propositions, and passes them to a remote executive, which then evaluates the assertion. The DBRover, NASA's Java Path Explorer, and NASA's Eagle are remote RV tools.

The code generated by the vanilla StateRover code generator is usually part of the application SUT. Nevertheless, statechart assertions can be also used within a *remote* monitor.

## 5.4.1. The Role of Assertions during REM

Recall from Section 5.1.4 that assertions are champions of a single concern and are therefore weaker than the model statechart being monitored. Stated differently, an assertion in its negative form (the form that flags failures) is *stronger* than the statechart; i.e., if an assertion evaluates to *false* for some input-sequence, output-sequence pair, the statechart being monitored should not generate that pair.

Indeed the role of an assertion in an REM environment is to find whether there are any input-sequence, output-sequence pairs that will cause it to evaluate to false, since whenever an assertion is flagged it means that the model and the assertion are possibly inconsistent. The following are reasons that might cause the REM tool to flag an assertion:

1. The statechart SUT does not comply with the requirement manifested by the assertion.

2. The assertion is incorrect—it does not capture the requirement properly—for any of the reasons listed in Section 5.3.3.

3. The input sequence violates a contract made with the model statechart. In that case, both the assertion and the model are correct, and the test is at fault.

### 5.4.2. Post-Mortem Verification vs. REM

With post-mortem verification, the program is executed and a history trace of the basic propositions pertaining to the LTL assertion (or assertions) being evaluated is stored. Later, when the program terminates, the LTL formula is evaluated using the stored trace as the input sequence.

Post-mortem verification is useful only for testing and cannot be used for runtime purposes [see the end of the chapter]. Another drawback is that the size of the storage is proportional to the length of the test. From a practical standpoint, post-mortem verification is weak because, once a violation of an assertion is discovered, the information cannot be *readily* used to debug.

## 5.5. Runtime Recovery from Requirement Violations

Imagine a robot, such as Nasa's Martian robot, that occasionally, when deployed, unexpectedly switches into reverse motion. The

obvious solution is to reset the robot, a process that obviously consumes precious time. Performing a reset is essentially the crudest
exception handling possible. Runtime recovery techniques suggest
performing finer exception handling and recovery when a temporal
assertion violation is discovered during runtime.

One type of runtime recovery technique, based on the TemporalRover code generator, uses two separate languages to specify the
assertion and the recovery mechanism. Assertions are specified in
LTL or MTL (or a combination of the two), and executed using code
generated by the TemporalRover. Recovery is specified using Java
exception handling and executed by the Java virtual machine (JVM).
This technique is therefore called a *two-level recovery*.

As an example, think of a Java program that controls a pump. A
method, *changeStatus(boolean b)*, turns the pump on and off, and
an LTL or MTL assertion is added to the *changeStatus* source code
as follows:

```
void changeStatus(boolean b) {

    m_status = b;

    /*TRBegin —- an embedded temporal logic assertion

        TRAssert{Always {m_status==OPEN} Implies

        Eventually_<30_{m_status==CLOSED}  } =>

        $throw new PumpException(); // throw an exception

                                       when the assertion

                                       fails

    TREnd*/

}
```

The Java code catches this exception wherever it fires the *change-
Status()* method, as in the following code:

```
try {

...

if (pressure < 10.0) changeStatus(OPEN);

if (pressure > 100.0) changeStatus(CLOSED);

} catch (PumpException e) {

    System.err.println("Pump Assertion #1 violation

    detected; closing pump...");

    m_status=CLOSED; // recovery!

}
```

With a one-level recovery, assertion execution and recovery are specified using the same specification language. The StateRover's recovery technique for statechart assertions is one-level, as illustrated in Figure 4.1. There, when Assertion1 fails, it reaches a terminal state (Figure 4.1b). The primary statechart (Figure 4.1a) then immediately jumps to the Green state, performing custom recovery in the primary statechart.

One-level recovery is clearly more readable than two-level. It is also simpler to specify, because all specification is done in UML using a single tool.

## 5.6. Automatic Test Generation

Automatic test generation (ATG) is precisely that. An ATG tool generates a huge number of test cases. Each is a sequence of inputs that is fed into the SUT, thereby exercising it.

With so many tests, a human tester cannot evaluate whether the SUT's behavior for each test is acceptable. First of all, she would not be able to, because she did not write the tests herself. Similarly, the sheer volume of tests makes it impossible for anyone to monitor the SUT's behavior for each test. ATG is therefore useful primarily when employed with automatic REM, as we will discuss below.

## 5.7. Execution-Based Model Checking

As we will see in this section, execution-based model checking is theoretically inferior to classical model checking, because execution-based model checking is automated testing combined with REM and therefore seldom yields 100% test coverage, whereas classical model checking consists of a mathematical proof that does yield 100% coverage. The truth, however, is that both classes of techniques require compromises. Execution-based model checking compromises in the achieved test coverage; classical compromises in the size and type of programs that can be verified, and in the kinds of assertions that can be verified to begin with.

More specifically, with execution-based model checking, we can verify interesting and meaningful assertions and properties using many interesting test cases, perhaps with some insight into the degree of test coverage actually achieved; classical model checking, on the other hand, provides total verification (100% coverage) of rather limited assertions and properties of programs of a limited size, possibly after inserting unknown errors as a result of a process called abstraction.

Obviously, if we have a SUT and assertions that can benefit from classical model checking, we should use it. Chances are, however, that the properties of interest and the size of the SUT will not facilitate the classical approach, and then execution-based model checking is the only option.

## 5.7.1. Classical Model Checking

Classical model checking, which we will generally refer to simply as *model checking,* consists of a program, or model, M, and an LTL assertion, $p$[1]. Assertions such as $p$ are typically associated with states of M and use propositions that assert about variables in M (e.g., $x > 0$) and the visitation of states in M (e.g., in-state-A, for state A of M). The grand idea behind this approach is to prove mathematically that M conforms to assertion $p$ using the negative interpretation we discussed in Section 5.1.4, that is, to prove that there is no input sequence, *in,* such that, when applied simultaneously to both M and $p$, causes M to generate an output sequence, *out,* such that $p$ fails on the pair *in, out.*

As we noted, a mathematical proof provides 100% coverage; in other words, if model checking says that $p$ is conformed to, no more testing is needed for that assertion. One major issue with 100% coverage is that most interesting models contain loops, such as the statechart model of Figure 2.7a. Consequently, the model can expect input sequences of any length. Typically, therefore, model checking (and other formal methods) consider LTL with an infinite-sequence semantics, discussed in Section 3.3.9.

The following is a simplified description of classical model checking:

1. M must be a $\omega$-DFA (discussed in Section 3.3.9). The process of converting a conventional program in C, C++, or Java into a reasonably small $\omega$-DFA, known as abstraction, is outside the scope of this book. It should be noted, however, that for most programs it involves throwing away some information and thereby reducing the number of states, that is,

---

[1]   Model checking often uses more complex specification languages, such as CTL, which we discussed briefly in Chapter 3.

the size of the ω-DFA to a manageable size. For example, a program with two 32-bit integers induces $2^{32*2}$ states in M, an unmanageable number. When a program results in such a large ω-DFA, it is said to be limited by the model-checking state explosion problem. Now, if the abstraction process (performed by a person or a program) concludes that only 4 bits of the 32 are actually used, then M is, in fact, much smaller ($2^{4*2}$).

2. The assertion $p$ is converted into a ω-DFA. This process induces a worst-case double exponential blowup; however, algorithms exist that tend to avoid the worst case most of the time. Also, LTL assertions tend to be rather small, so this problem is typically dwarfed by the ω-DFA state-explosion problem.

3. Model checking is now reduced to the automata-theoretic problem: is there a string accepted by M's ω-DFA that is not accepted by p's ω-DFA?

The primary drawbacks of classical model checking are:

1. State explosion. Dealing with this problem requires very efficient abstraction techniques. But those techniques pose a new problem for verification, which is: *haven't we thrown out the baby with the bath water?* In other words, is it possible that abstraction introduced artifacts that hide errors? Say, for example, that an abstractor concludes that all behavior of a 32-bit integer variable, $x$, is captured by the 4 least significant bits (lsb's), perhaps because the abstractor reasons that the program assigns values only in the range of 0 through 15 to $x$. Now say that the user wrote a buggy increment operation modulus 16. As a result, after abstraction, model checking would miss the bug and announce a successful behavior

because it is looking at just 4 lsb's alone, but it would not miss the bug if it considered the entire 32-bit space.

Nevertheless, more effective model-checking techniques have been developed recently that address the state-explosion problem with certain degrees of success, such as the use of SAT solvers for model checking.

2. The specification language problem can be described as follows: the kinds of properties, or assertions, one can verify with model checkers are limited to begin with. For example, one primary issue is the lack of real-time or time-series constraints within the specification of requirements that model checkers can handle. Model checking typically works for plain vanilla temporal logic such as LTL, but does not support the specification of real-time constraints or time-series constraints. That essentially rules out its use for the verification of practical reactive systems, because assertions for real-life systems almost always have such constraints. In addition, temporal logic, the language of choice for most model checkers, suffers from many other drawbacks, as we discussed in Chapter 3.

## 5.7.2. Execution-Based Model Checking

To repeat, execution-based model checking combines ATG on one side, and an REM tool on the other, as depicted in Figure 5.1. The ATG side is responsible for generating a large volume of test cases, each with its own sequence of inputs that exercise the SUT. The REM side then monitors both the SUT's inputs and its outputs and decides whether to accept or reject each behavior.

Execution-based model-checking techniques come in two primary flavors: black-box and white-box.

### 5.7.3. Black-Box Execution-Based Model Checking

Black-box execution-based model checking consists of a special program, called the *environment model,* constructed by a human tester who is responsible for creating input sequences to the SUT. The environment model is essentially a glorified automatic test generation tool: instead of writing one test (one scenario) at a time, a human tester writes a program that generates a large set of tests. It does so by mimicking certain situations or contexts in the SUT's environment. For example, a Java environment model for a traffic light controller might mimic cars driving down the main road at a rate of $x$ cars every 10 seconds, using a stochastic choice of $x$ representing one, two, or three cars. Alternatively, the tester could use statecharts to write a probabilistic environment model for a TLC, as we considered in Section 2.11.

NASA's Java Path Finder (JPF) is a black-box test generator that works for Java only. Its environment model consists of a special extension of Java. The combined model (the Java SUT and the environment model) must be a closed system, that is, every possible external input to and output from the SUT must be driven from or to the environment model and vice-versa.

### 5.7.4. White-Box Execution-Based Model Checking

As its name suggests, an automatic white-box test generator is an ATG tool that observes the SUT, the specification assertions, or both, and generates tests based on that information. A white-box test generator is said to be *model-based* when it observes the SUT only, and specification-based when it observes the assertions only. The ATGRover, for example, is a specification-based white-box test generator that automatically creates tests for a program annotated with LTL assertions (à la the TemporalRover, which we discussed in Chapter 3). It attempts to create tests that fail those LTL assertions. The StateRover's white-box test generator, in contrast, is a hybrid; that is, it is both model-based and specification-based.

**The StateRover's White-Box Test Generator**

The StateRover's white-box test generator is an automatic code generator for a JUnit TestCase class. This TestCase, however, does not capture a single, manually created scenario, as JUnit test cases usually do. Rather, it contains a loop that creates a plurality of tests for a SUT statechart. Class TestCase1.java in the example in the Appendix contains the white-box test generator test-unit method created by the StateRover for the example's statechart SUT. This test-case is denoted the *WBTestCase*.

Automatically generated tests are used in three ways:

1. To search for severe programming errors of the kind that induce JUnit error states, such as *NullPointerException*.

2. To identify tests that violate temporal assertions. Such failed assertions should be captured by a JUnit *assertFalse()* statement, using the *isSuccess()* feedback loop depicted in Figure 4.2.

3. To identify input sequences that lead the SUT-statechart to particular states of interest.

The *WBTestCase* creates sequences of events and conditions for the SUT. As we have said, a SUT-statechart that contains loops, which is the most common kind of SUT, has an infinite number of input sequences and input sequences of any length. It is therefore impossible to create the entire test universe. The StateRover white-box test generator addresses these issues in the following way:

1. The user specifies the maximum number of test sequences the WBTestCase is allowed to generate, denoted as the *WB test-budget*.

2. The user specifies the maximum length of any test sequence generated by the WBTestCase.

The WBTestCase is intelligent in the following regard: it creates only sequences which "matter" either to the SUT or to an assertion statechart. For example, consider the following 3-cycle long test generation episode for the SUT in Figure 5.2 (also found in Example 1 of the Appendix).

**E5.7.4.1**:

**Cycle-1**: Upon start-up, the SUT has only one option: to observe the *start* event. The WBTestCase therefore generates this event. Consequently, the SUT moves to the *Green* state.

**Cycle-2**: When in the *Green* state, the SUT expects either a *time-out* event or a *reset* event, so the WBTestCase generates one of those, using one of two methods that the user selects: a *stochastic* method or a *deterministic* method, which we will discuss below. Say the WBTestCase decides to generate the *timeout* event; the SUT will then transition to the *Red* coarse state, specifically to the *On* and *Count_0* states.

**Cycle-3**: There, the SUT expects one of the following events: *newCar*, *timeout*, and *reset*. As in Cycle-2, the WBTestCase decides to generate one of those.

Hence, in general, the StateRover's WBTestCase repeatedly observes all events that can possibly occur when the SUT is in a given state configuration, generates one event and fires the SUT using that event. This process continues until the user-specified test-length is reached; this concludes a single test generated by the WBTestCase.

It then generates another test. The next test scenario should, if possible, differ somehow from the first scenario. This is where the two methods mentioned earlier differ.

The deterministic method generates scenarios for E5.7.4.1 in an orderly manner, such as:

Scenario #1: *start.timeout.newCar*

Scenario #2: *start.timeout.timeout*

Scenario #3: *start.timeout.reset*

and so forth.

The stochastic method generates scenarios for E5.7.4.1 using a randomized method in each cycle. For example, in cycle #2 it tosses a coin to decide between the *timeout* and *reset* event generation.

Note that the stochastic method is not a purely random method; it is in fact as intelligent as the deterministic method. Both methods generate a smart set of events that "matter" on a cycle-by-cycle basis. The only difference between the two methods is in the order in which the events are generated. The stochastic method picks events from the smart set using a random selection approach, whereas the deterministic method uses a fixed order for doing so.

This process is the model-based aspect of the StateRover's white-box test generator. However, the white-box test generator actually observes *all* entities, namely the SUT and all embedded assertions. It collects all possible events from all those entities, thus creating the specification-based component of the tool. For example, in cycle #1 of E5.7.4.1 there are now two events that can affect the participating statecharts: *start*—which affects the SUT, and *newCar*—which affects the assertion Assertion_NewCar_Contract.

## White-Box Test Generator of Time and Data

Up to this point, we have looked at the process of generating intelligent sequences of *events* for the SUT. However, other artifacts potentially affect statechart SUT behavior as well:

1. Condition guards, as in cycle #2 of E5.7.4.1 where the Boolean variables *manualOn* and *insideJunction* affect the control flow of the SUT while in state *On*.

2. Data-objects passed as arguments to event handlers, like *newCar(Car car, Rolls rolls)* in Figure 2.17.

3. Timing information for *timeoutFire* events.

The StateRover white-box test generator does not generate condition guard information. It is assumed that this information is a derivative of a computation in an external component.

For data-objects the StateRover white-box test generator integrates with a Java object-factory, supplied by the user. The object-factory is any class that implements the ITRAtgObjects interface with the methods:

1. *int getMaxPossibleObjects(String sType),* that returns the maximum number of objects of a given type that the object-factory might create. For example:

```
public int getMaxPossibleObjects(String sType) {

    if (sType.equals("Car")) return 2;

    if (sType.equals("Rolls")) return 3;

    return 1;

}
```

2. *Object getObject(String sType, int n)*, that returns object number *n* of a given type (*n* is a number between 0 and the maximum provided by the first interface method). For example:

```
public Object getObject(String sType, int n) {

    if (sType.equals("Car")) return new Car();

    if (sType.equals("Rolls")) return new Rolls();

    return null;

}
```

It is the object-factory's responsibility to generate new objects for the white-box test generator.

The white-box test generator generates *timeoutFire* events using two possible approaches. The first approach, used by the deterministic method, treats this event like any other external event. That does not represent timer events accurately, because if two timers, one for 10 seconds and one for one minute, start measuring time simultaneously, clearly the 10-second *timeoutFire* will fire first. In fact, since the 10-second interval fits six times within the one-minute interval, we would expect repeated invocations of these *timeoutFire* events to reflect that fact.

The second approach, used by the stochastic method, is more accurate. It generates increments of simulated time. When the increments pass a certain timer limit, the corresponding *timeoutFire* event fires.

### White-Box Test Generator Loop Management

Both types of *WBTestCase* address SUT-loops by allowing the user to specify a limit on the number of times a *WBTestCase* can force the SUT to visit the same state configuration. Thus, the deterministic *WBTestCase* is able to actually provide some coverage information. For example, if the user specifies that the SUT is permitted to visit a state-configuration at most once, which amounts to exploring simple paths in the SUT-statechart, it is quite possible that the *WBTestCase* would generate all such tests within its test-budget; the deterministic *WBTestCase* would then announce that it has covered all *simple-paths*. If it is unable to cover all possible paths of a given maximum length within its test-budget, the deterministic *WBTestCase* would attempt to estimate what percentage of tests it actually generated.

**The White-Box Test Generator and Input Assertions**

Given sufficient time, the white-box test generator will generate a test that violates an input contract assertion if such a contract exists. For example, at some point the white-box test generator in the Appendix will generate a test sequence that violates *Assertion_NewCar_Contract*. Such a violation by no means indicates that the SUT is faulty. Rather, it is a case in which the SUT is exercised by a scenario that it is rightfully not expecting. Therefore, the StateRover white-box test generator actually ignores SUT failures when the generated test scenario violates an input assertion.

**Debugging Automatically Generated Tests**

When an automatically generated test T violates an assertion, it is likely that the assertion developer or tester would like to rerun that test alone, possibly in the context of fewer assertions (those that failed). To that end, rather than manually writing a JUnit test-Case that repeats the sequences of T, the white-box test generator

*WBTestCase* can be executed in a *single test mode*, using a unique ID that identifies T.

**The White-Box Test Generator and Assertion Coverage**

Using a white-box test generator automates the test generation process, but it requires an automatic observer on the output side. We want to rely on assertion monitoring to perform the automatic observation. For argument's sake, let's say we run a test suite using an automatically generated *WBTestCase* on a SUT that contains no assertions. The *WBTestCase* will then succeed in all tests because there is no assertion to fail; this is the extreme example of poor *assertion coverage*, which is a numeric indication of the suitability of the assertions for the underlying SUT.

The StateRover white-box test generator estimates assertion coverage using the notion of an assertion being *touched* by an event, which holds true if the assertion traverses a transition as a result of the event firing. Consequently the StateRover white-box test generator provides the following assertion coverage information:

1.  The ratio of cycles in which the assertion was touched, to all cycles.

2.  The ratio of tests, or runs, in which the assertion was touched, to all runs.

## 5.7.5. White-Box Execution-Based Model Checking = White-Box Test Generation + REM

In the literature, we find that execution-based model checking is defined as *a technique for exploring the possible execution runs of a program by executing the program repeatedly in a systematic manner while comparing its behavior to the behavior specified by formal specification assertions.*

This is in fact the outcome of our complete automated testing architecture illustrated in Figure 5.1. Our automatic white-box test generator wrapped in a JUnit TestCase is responsible for "executing the program repeatedly in a systematic manner," and a REM tool is responsible for comparing the program's behavior to "the behavior specified by formal specification assertions."

# Chapter 6

## Application of Formal Specifications and Runtime Monitoring to the Ballistic Missile Defense Project

*By Nick Sklavounos*

## Introduction

Nick Sklavounos is and has been a key modeler and developer within the Missile Defense National Team (MDNT), the collaboration of companies across the defense industry responsible for the Missile Defense project. In this capacity Nick has witnessed the conception of the ideas described in this book. In the early days of this project the primary production-level tools that were commercially available for formal specification and run-time monitoring were LTL- and MTL-based. Nick and his group started of writing and simulating assertions in LTL and MTL using the DBRover tool. The considerable feedback the MDNT provided me helped developed the concept of *verifying statecharts using statecharts*, which is what this book is about.

In this chapter Nick describes the process his team has and is using to develop assertions and also describes some of the assertions used so far.

The observant reader will notice that the assertions Nick describes in this chapter are rather simple. This fact had an important effect on the decision to use statecharts for writing statecharts because users resisted using a heavyweight tool like the DBRover and Temporal-Rover to specify, monitor, and perhaps even recover from failures of simple requirements. Hence they are now modeled as simple deterministic statecharts.

Nevertheless, the team has since developed more complex assertions such as nondeterministic assertions that use quantification.

## 6.1. Abstract

As the Missile Defense Agency (MDA) evolves a globally integrated missile defense system-of-systems, software will continue to play an integral role within the automation of planning, command and control (C2), and battle management (BM) functions. With an increased reliance on software decision aids to support mission objectives, the topic of software reliability has received much attention across the Missile Defense community. Process reengineering efforts have focused upon the integration of formal specification and verification techniques in order to build additional reliability within systems under development. Enforcing the necessary rigor within specification and verification techniques, however, poses a significant challenge for Department of Defense (DoD) software development efforts, specifically in light of recent trends toward light(er)-weight, spiral and agile software lifecycle strategies. The following section will discuss hands-on experience and lessons learned from applying a lightweight formal specification and runtime monitoring approach to the Ballistic Missile Defense Project.

## 6.2. Context

I will begin by providing the reader some background context in regard to the effort's objectives, organization, and development environment. A dedicated team of software engineers and developers assembled with the following two objectives:

1. To engineer a software development methodology that results in more reliable software;

2. To pilot the proposed software development methodology within a realistic development environment.

Formal specification and verification techniques were explored and integrated in order to achieve additional reliability within the code base. The effort was also unique in that it embraced model-based development within an agile process. The process pilot involved the development of sensor tracking software, which represented a narrow slice of the envisioned overarching Battle Management Command and Control (BMC2) architecture. Additional relevant characteristics of the domain were the following:

1. A transformational architecture, with mathematically intensive algorithms;

2. A sparse set of formally documented requirements, specifically within the area of performance (i.e. timing, reliability).

## 6.3. Formal Specification and Verification Approach

A special team was devoted to the definition, execution, and management of a lightweight formal specification and runtime monitoring approach. The approach employed was the following:

1. Define a set of specifications within natural language.

2. Express the specifications formally, within the form of a Harel statechart.

3. Simulate the formal specification, and verify proper statechart behavior.

4. Instrument the System-Under-Test[1] (SUT) in order to publish metrics (events) to the formal specification statechart.

5. Stimulate the System-Under-Test and monitor the formal specifications.

6. Analyze discrepancies among the implementation and formal specifications, and refine system and/or specifications accordingly.

The following three tool sets were employed in order to support the formal specification and verification approach:

1. **StateRover** *(Time-Rover Inc.):* for model-generated formal specification code;

2. **Eclipse**: for JAVA development within an integrated development environment (IDE);

3. **JUNIT***:* for unit testing with the formal specifications.

The following subsections will describe details, processes, and strategies that accompany each of the steps within the formal specification and verification approach.

### 1. *Define a set of specifications within natural language.*

Identifying a complete and correct set of specifications is an iterative, time-consuming, and often analysis-intensive activity.

---

[1] Please note that for clarification purposes, the author uses the term "Statechart-Under-Test" throughout the book; however, the author of this chapter prefers the term "System-Under-Test."

For domains that are highly volatile and/or not well understood, deriving specifications is an even larger challenge. In Department of Defense (DoD) software development environments, large-scale modeling and simulation activities are often tailored toward deriving specifications. These efforts tend to be time-consuming, and require dedicated subject matter experts, lab support, and analysts. Due to resource, timing, and schedule constraints, a lightweight technique was employed in order to define an initial set of specifications. The goals for defining a set of natural language specifications were the following:

1. To derive a "meaningful" set of initial specifications. Specifically, to derive the specifications from a structured reliability analysis activity in which system level failures are identified, along with their associated safety-critical controlling strategies.

2. To focus on specifying safety properties (what the system should not do), in addition to liveness behaviors (what the system should do). Specifically, to focus on safety, reliability, and timing specifications, since these specifications were not thoroughly resident within the standard requirements documentation.

In order to realize the stated objectives, the natural language specifications were derived from a reliability analysis activity, based upon concepts from John D. Musa's *Software Reliability Engineering* process. In order to better understand the activity, I'll begin by clarifying some key definitions. The definitions of failures, faults, and formal specifications are stated below:

*Failure:* Departure of a system behavior in execution from user's needs. (Musa, *Software Reliability Engineering*)

*Fault:* Defect in system implementation that causes the failure when executed. (Musa, *Software Reliability Engineering*)

Thus, a failure is a manifestation of faults and is user-oriented (as opposed to developer-oriented). We'll define the term formal specification as follows:

***Formal Specification:***   A machine-interpretable description of safety and/or liveness properties within the system-under-test. Each specification becomes a runtime check within the system-under-test that detects and controls faults, thus minimizing the risk of system level failures.

Figure 6.1 below depicts the logical relations among failures, faults, and formal specifications.

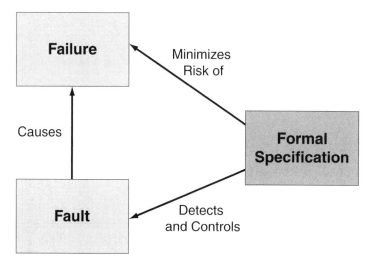

1. Faults potentially cause system failures.
2. Formal specifications are runtime checks that detect and control faults, thus minimizing the risk of system failures.

FIGURE 6.1   The relations among failures, faults, and formal specifications.

Thus, the approach employed was characterized by identifying failures, identifying their potential causes, and defining control mechanisms (checkable properties) and recovery strategies. The process employed is depicted within Figure 6.2:

1. Define a set of System-Level Failures.
2. Determine potential causes for System-Level Failures.
3. Define checkable properties in which to detect potential failure causes (faults), and capture as natural language specifications;
   a. Define recovery strategies, if applicable, for each specification (for when the specification is not met).
4. Document all failures, causes, natural language specifications, and recovery strategies within a Failure Table.

FIGURE 6.2   Deriving natural language specifications from a failure analysis.

The resulting output from the process is a failure table, depicting the traceability among failures, causes (faults), specifications, and recovery strategies. A sample failure table is depicted within Figure 6.3.

| Failure Description | Potential Cause (Fault) | Natural Language Specification | Recovery Strategy |
|---|---|---|---|
| System does not engage threat in time. | System does not assign a weapon system to a threat in enough time. | "When a threat is identified, a weapon shall be paired with the threat within x seconds." | Log & notify operator |
| System does not engage threat successfully. | System assigns a weapon that is not capable of engaging the threat. | "When an engagement opportunity is identified, never shall a weapon be assigned prior to checking whether that weapon system is capable of engaging the threat." | Check operational status and intervene accordingly. |
| System incorrectly labels a threat as a nonthreat | System discards too many valid tracks. | "When tracking process begins, never shall more than x tracks be discarded prior to completion." | Roll-back and re-examine discarded tracks. |
| System is not available. | System is overloaded. | "When tracking process begins, never shall fewer than x health and status updates be reported every y seconds." | Spawn additional tracking process. |

FIGURE 6.3   Sample failure table.

In order to derive additional natural language specifications, a "white-box" approach was employed in which system design artifacts were reviewed. A structured process was employed that involved traversing Unified Modeling Language (UML) artifacts, and working with Subject Matter Experts (SMEs) in order to answer a set of specific questions. The questions were tailored toward capturing unspecified system behaviors. The activity was labeled "armor-plating" of statecharts and activity diagrams, since the derived specifications served as an additional layer of behavioral requirements, in addition to those captured with traditional UML diagram semantics). The following sample templates, depicted within Figures 6.4, 6.5, and 6.6, demonstrate a technique for "armor-plating" of UML statecharts and activity diagrams.

---

**Armor Plating a State Chart Diagram**

1. Specify sequences of transitions that are unacceptable.

2. Specify sequences of states that are unacceptable.

3. Specify critical regions (states that cannot be accessed concurrently).

4. Specify states that cannot be visited repeated times.
   a. Specify states that cannot be visited repeated times within a timeframe.

5. Specify states that cannot be visited repeated times sequentially.
   a. Specify states that cannot be visited repeated times sequentially within a timeframe.

6. Specify events that cannot fire repeated times.
   a. Specify events that cannot fire repeated times within a timeframe.

7. Specify events that cannot fire repeated times sequentially.
   a. Specify events that cannot fire repeated times sequentially within a timeframe.

---

FIGURE 6.4   Armor-plating of UML statechart diagrams.

---

**Armor Plating an Activity Diagram**

For each activity:

1. Specify unacceptable inputs.

2. Specify unacceptable outputs.

3. Specify unacceptable input/output pairs.

4. Specify unacceptable conditions (invariants) throughout activity execution.

5. Specify unacceptable time duration.

6. Specify unnacceptable average duration (across multiple runs).

7. Specify earliest/latest start time.

8. Specify earliest/latest finish time.

---

FIGURE 6.5   Armor-plating of UML activity diagrams.

---

**1. Specify recovery strategies for each specification, if applicable.**

A. Specify a recovery strategy if the specification is not met.

B. Specify a recovery strategy if the specification is met *once*.

C. Specify a recovery strategy if the specification is not met *multiple times*.

D. Specify timing constraints for each recovery strategy.

---

FIGURE 6.6   Specifying recovery strategies.

The specifications from each of the derivation exercises were maintained within a configuration managed repository, and revisited on an iterative basis.

## 2. *Express the assertion formally, within the form of a Harel Statechart.*

In order to create machine-readable specifications to verify at runtime, each of the natural language specifications were expressed with formal semantics. Harel statechart semantics were employed and enforced by the StateRover modeling tool. In order to aid in expressing the natural language specifications formally, a library of formal specification patterns was defined. The library identified common natural language specification requirements, and their associated statechart representations. The library was maintained throughout development and used as a "look-up" tool in order to translate a natural language specification to a formal representation. A subset of the pattern library is depicted below within Figure 6.7. Sample formal specification statecharts are depicted within Figures 6.8, 6.9, 6.10, and 6.11. Formal Specification statecharts will be discussed in details within the subsequent section.

| Pattern Name | Natural Language | Examples | State Chart Pattern |
|---|---|---|---|
| P1: Timing of Events | "When A, then eventually B within x seconds." | "When a threat is identified, a weapon shall be paired with the threat within x seconds." | S1 |
| P2: Precedence of Events | "When A, then never B prior to C." | "When an engagement opportunity is identified, never shall a weapon be assigned prior to checking whether that weapon system is capable of engaging the threat." | S2 |
| P3: Counting of Events | "When A, then never B more than x times prior to C." | "When tracking process begins, never shall more than x tracks be discarded prior to completion." | S3 |
| P4: Counting of Events Over Time | "When A, then never B fewer than x times every y seconds." | "When tracking process begins, never shall fewer than x health and status updates be reported every y seconds." | S4 |

FIGURE 6.7   Formal specification pattern library.

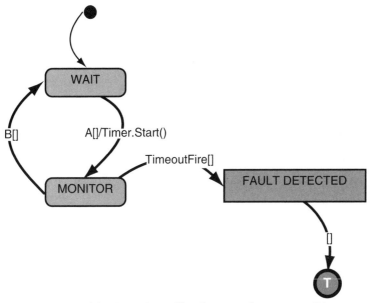

FIGURE 6.8   Formal specification statechart pattern S1.

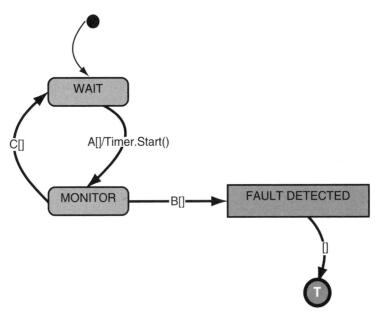

FIGURE 6.9   Formal specification statechart pattern S2.

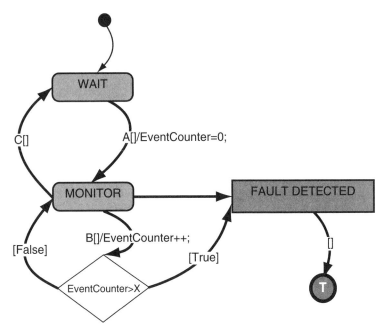

FIGURE 6.10   Formal specification statechart pattern S3.

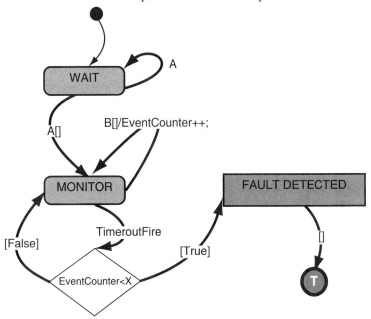

FIGURE 6.11   Formal specification statechart pattern S4.

## 6.3.1. Formal Specification Statecharts

Each formal specification statechart is comprised of events, states, activities, local variables, and action code. Events define stimulus in which the formal specification statechart reacts accordingly. For example, statechart pattern S2 (Figure 6.9) is comprised of three events, A, B, and C. Each of the formal specification statecharts has a minimum of two states, labeled "WAIT" and "MONITOR." Upon construction or reset, each statechart waits within the "WAIT" state until a relevant event fires. Upon event firing, the statechart will transition to a "MONITOR" state. This indicates that the formal specification is active, and is under evaluation based upon external events (either from the system-under-test or from a timer). The statecharts are also comprised of local variables, which are employed in order to specify clock parameters, counters, or any other variables necessary for policy-checking computations. A formal specification statechart can also possess a clock object, which can be passed a timing threshold. The timing threshold indicates when the timer object should fire the *timeoutFire()* event on the formal specification statechart. Lastly, action code can also be embedded within states, transitions, and activities (not depicted within Figures 6.8–6.11). The action code can be employed in order to call external computation methods, or simply to log to the console. The following section will describe the verification technique employed for formal specification statechart behavior.

3. *Simulate the formal specification, and verify proper behavior.*

In order to verify the proper behavior of the formal specification statecharts, unit tests were implemented (within the JUnit framework) for each statechart pattern. Figures 6.12 and 6.13 depict two notional test cases for formal specification statechart pattern S1.

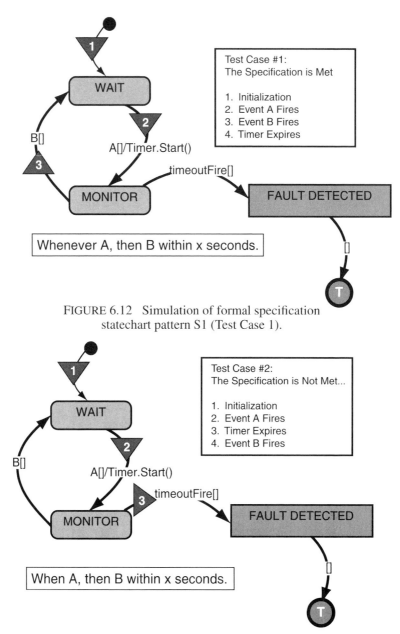

FIGURE 6.12   Simulation of formal specification
statechart pattern S1 (Test Case 1).

FIGURE 6.13   Simulation of formal specification
statechart pattern S1 (Test Case 2).

4. ***Instrument the System Under Test (SUT) in order to publish metrics (events) to the formal specification statechart.***

Once the behavior of the formal specification statecharts was verified, then the system-under-test was instrumented in order to report metrics (events) to the formal specification statecharts. The developers placed "hooks" within the code base, in order to fire the relevant events on the formal specification statecharts.

An alternate approach is the use the StateRover tool in order to instrument the system-under-test. The StateRover tool supports an integrated statechart model, which depicts both the system-under-test statechart along with the formal specification statechart. Within the tool, the developer can then map events from the system-under-test statechart to events within the formal specification statechart. Thus, when code is generated for both statechart models, the tool will enforce that relevant statechart events are fired for both the system-under-test and the formal specification statechart models. The employment of the StateRover tool also supports modeling of recovery strategies in which behaviors can be executed when a formal specification is not met (specifically when the statechart reaches a "FAULT DETECTED" state). Lastly, a third potential architectural instrumentation approach involves publishing events to a centralized data repository. The formal specification statecharts can then query the centralized repository (possibly an event queue) and process the events accordingly.

5. ***Stimulate the system under test and monitor the formal specifications.***

6. ***Analyze discrepancies among the implementation and formal specifications, and refine system and/or specifications accordingly.***

Once the instrumentation was complete, then testing with formal specifications began. The system was stimulated, and the formal specifications were able to detect and control faults. The described architecture is depicted within Figure 6.14 below.

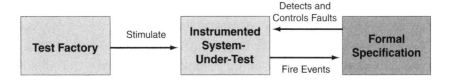

1. The **Test Factory** stimulates the **System-Under-Test**.

2. The **Instrumented System-Under-Test** fires events on the **Formal Specification** state charts.

3. The **Formal Specification** state charts detect and control faults within the **System-Under-Test**.

FIGURE 6.14

The formal specifications were monitored during both unit and system level test activities, and results were captured within test reports. On test runs, the parameters within the formal specification statecharts were varied in order to determine performance sensitivities. When the formal specifications were not adhered to on test runs, a root cause analysis was performed in order to determine whether the fault was attributable to the formal specifications themselves, or the system-under-test. The employment of formal specification statecharts during the test phase did not replace complimentary standard test processes, but served as an additional data point in order to support unit and system level test activities.

## 6.4. Overall Value

The formal specification and verification process provided additional value specifically within the area of requirements engineering. Capturing the specifications with formal semantics resulted in "tighter" behavioral descriptions, removing many of the ambiguities within natural language descriptions. Further, the process demonstrated value in complimenting the traditional requirements flow-down process, by feeding additional requirements back into the specification. Many of these additional requirements took the form of negative specifications and timing specifications, and thus provided additional perspectives of the system's behavior.

In terms of process, the employment of the pattern library to capture reusable formal specification patterns was beneficial in providing the Subject Matter Experts (SMEs) insights as to commonly employed specifications, and also to provide reusable software components in order to reduce overall statechart modeling efforts. The visual statechart models are commonly known and understood by developers, and served as a more practical solution than employment of a high-level specification language such as Linear Temporal Logic (LTL). The technique of deriving specifications from system-level failures proved useful in ensuring that critical specifications were captured initially. Further, it supported resource allocation of test and developer resources to formal specifications work on a cycle-by-cycle basis. Demonstrating the relationship among faults and failures, and also the relationship of runtime-checking to fault controlling, aided in building a stronger case in order to justify the allocation of test and developer resources to formal specifications work.

From a verification standpoint, implementing the specifications with machine-readable, formal semantics enabled the automation of specification checking. This served as a valuable complimentary activity to traditional unit testing. Further, the formal specifications could be left in the code base to support both unit and system testing. The practice of varying the parameters within the formal specifications (during test time) also served as a valuable tool in order to examine the system's performance sensitivities.

## 6.5. Challenges

The pilot demonstrated that a number of programmatic and technical challenges must be overcome in order to successfully integrate a formal specification and verification process within a large scale software development effort. Resource allocation must account for training in formal specification patterns, modeling techniques, tooling, and runtime monitoring. Subject Matter Experts (SMEs) must maintain a flexible mind-frame, and be willing to pursue nontraditional specification strategies. Further, developers must be devoted to implementing the formal specifications, and testers must recognize and utilize the formal specifications that exist within the System Under Test (SUT).

Some strategies that were effective in keeping a focus on formal specifications included selection of a specific subset of formal specifications to implement per development iteration, and also allocation of a separate development team in order to model and implement the formal specifications. The StateRover tool supports such a model, in that it allows for concurrent development of the System Under Test (SUT) and the formal specifications (decoupling through an interface definition). These methods proved adequate for a small development team, but within larger development environments, coordination and management will pose substantial challenges. Future experimentation is necessary in order to demonstrate whether the formal specification and verification process will efficiently scale up to large software development efforts. Finally, other future challenges include the alignment of formal specification processes with standards, configuration management of formal specifications, the employment of formal specifications for reliability quantification, and the evolution of supporting tool sets.

# Appendix

# TLCharts: Syntax and Semantics

## A.1. About TLCharts

The entry example consists of the following four conditions: (system) *begin*, (system) *end*, *keyPressed*, *doorOpen* (where *doorClosed* = ¬*doorOpen*) and *alarm*. A particular requirement of interest is:

R: a session is the interval between a *begin* and an *end-condition*. For every such session a *keyPressed* must be repeatedly sensed within two-minute intervals or else an *alarm* must sound within 10 seconds until *keyPressed* is sensed. Also according to this specification, once the alarm sounds then the assertion has succeeded and no more alarms are permitted. The *end-condition* is defined to be true whenever there is an interval which starts with the door closed followed by an *end* being repeatedly sensed until a later time when *begin* is sensed.

We refer the reader to the infusion pump example of [D3] which consists of a requirement similar to R. This reference also contains an LTL/MTL specification for the infusion pump requirement and analyzes subtle specification errors that result from the use of linear time temporal logic. A deterministic Harel statechart specification of requirement R is illustrated in Figure A.1. The discussion in [D3] compares the accuracy of the statechart of Figure A.1 with that of the TLChart of Figure A.2.

Figure A.1 and Figure A.2 are both legal TLCharts, i.e., Harel statecharts are a special case of TLCharts, and so are LTL and MTL assertions (using a diagram with a single state and transitions annotated with LTL).

TLChart extend Harel statecharts in the following ways:

1. Some transitions are annotated with LTL, MTL or TLS conditions, such as the transition labeled *alarm U keyPressed* in Figure A.2.

2. TLChart's support nondeterministic behavior.

3. TLCharts flavor of nondeterminism incorporates the specification of both good and bad computations with ambiguities resolved via a priority-based resolution scheme. From an automata theoretic perspective this amounts to existential nondeterminism with negation.

A TLChart input string represents a sequence of combinations of stimuli and corresponding system responses; for example, a sequence may contain *keyPressed*—generated by the environment, as well as alarm—a system generated response. The TLChart of Figure A.2 describes legal (accepted) and illegal (rejected) sequences. From a verification standpoint, a rejected string means that the systems behavior does not comply with the specification, typically due to an incorrect system reaction to the input stimuli. This application of diagrams to specification rather than programming and design explains the existence of a sink state (the Error state), which does not typically exist in a design phase statechart.

## A.2. Syntax

In this paper we consider Harel statecharts as first described by David Harel, including state hierarchy, concurrence, and history states. Hence, no state overlapping is permitted; this assumption will be changed in the next section. For simplicity, we assume that statechart transitions are annotated with conditions and no events, although we expect TLChart to be used and applied with events and conditions, much like UML statecharts. We use the automata-theory-oriented notation where transitions are annotated with symbols from a finite input alphabet $\Sigma$. A practical generalization of the automata model is to use the power set of $\Sigma$ as the actual input alphabet; this generalization enables multiple simultaneous input channels to the statechart device. Consequently, transition labels are subsets of $\Sigma$, where a transition labeled *label* means informally that all symbols $s \in$ *label* must be concurrently present on the input tape for the transition to be traversed.

TLChart transitions are annotated with one or both of the following types of conditions: *propositional* and *temporal*. Propositional conditions are subsets of $\Sigma$. Temporal conditions include all legal LTL and MTL formulae. In Figure A.2 temporal conditions are represented using curly braces. Hence [*end* {*end U begin*}] represents the propositional condition *end* and the temporal condition *end U begin*. When using (extended) regular expressions instead of LTL the equivalent condition is [*end* {*end\* begin*}].

A TLChart without state overlapping induces an *and/or* state tree as illustrated in Figure A.3. A TLChart with overlapping states, such as the TLChart of Figure A.4, induces an *and/or* state Directed Acyclic Graph (DAG), as illustrated in Figure A.5.

## A.3. Semantics without Temporal Conditions

The TLChart formalism specifies requirements using formal languages. The semantics of a TLChart is interlingua-based, using an *Equivalent Nondeterministic Automaton* (ENFA) Once defined in terms of its ENFA, a TLChart defines correctness properties in a manner that resembles formal logic specification, such as temporal logic specification. It observes a given input tape and decides whether this tape is acceptable or not. In real life terms the input tape corresponds to a combined sequence of inputs to-, and manifested outputs from-, a given system.

**Definition 1**. A *configuration* $C$ is a subset of the TLChart state set in which states of $C$ are pairwise orthogonal, i.e., if for every pair $r, s \in C$, the least common ancestor of $r$ and $s$ in the state tree is an *and* state.

**Definition 2.** A configuration $C$ is *maximal* if it is not a subset of any other configuration with a larger cardinality.

For example, in Figure A.2, the configuration {*Wait-For-Key-Pressed*} is not maximal while {*Wait-For-KeyPressed, Closed*} and {*Wait-For-KeyPressed, Open*} are maximal. Note that state configurations do not, in general, contain information about corresponding superstates, such as *Wait-For-KeyPressed* and *Closed* residing under *State-2*, which in turn resides under *State-1*. This information is not necessary because, absent state overlapping, state hierarchy is unique. However, we will change this notation when we describe TLCharts with overlapping states.

The ENFA's state set consists of all possible maximal configurations. Intuitively, this set represents a Cartesian product of state sets of concurrent TLChart threads. We denote a TLChart state within a configuration $C$ as a *constituent* of $C$.

As a preliminary step, before we describe the ENFA's transition relation, note that we can replace statechart and TLChart hierarchical transitions, such as *State-1→Init* in Figure A.2, with concurrence, using a new concurrent thread with one inner state, e.g. *State-1a*. The hierarchical transition is then replaced with the transition *State-1a→Init*. Also, in the following discussion we represent TLChart transitions as annotated binary relations over configurations. For example, in Figure A.1, the hyper edged transition labeled *begin* is the annotated relationship $(\{CLS, State\text{-}4\}, \{Init\})_{begin}$ and, in Figure A.2, the transition *Closed → Init* is represented by the annotated relationship $(\{Closed\}, \{Init\})_{end, \, end \, U \, begin}$.

For simplicity, we will first define consider TLCharts with no temporal conditions

**Definition 3.** A *global transition* is an annotated binary relationship between configurations.

**Definition 4.** Given a TLChart transition $(b',c')_{a'}$, we say it is a *member TLChart-transition* of a global transition $(b,c)_a$ if $b'\subseteq b$, $c'\subseteq c$, and $a'\subseteq a$.

Given a global transition $t=(b,c)_a$, we denote its set of member TLChart transitions as $m(t)$. For example, in Figure A.1, $t = (\{OPN, State\text{-}3\}, \{CLS, State\text{-}4\}_{\{doorClosed, \, end\}}$ is a global transition, where $OPN\to_{doorClosed} CLS$ and $State\text{-}3\to_{end} State\text{-}4$ are two members of - $t$.

**Definition 5.** Let $(b',c')_{a'}$ be a member transition of some transition $(b,c)_a$, then every state in $b'$ $(c')$ is said to be a *source* (*target*) state in $b$ $(c)$, and we denote the set of all source (target) states for $(b,c)_a$ as the *source* (*target*) *set* of $(b,c)_a$.

Note that member TLChart transitions of a global transition $t=(b,c)_a$ are always nonconflicting otherwise $c$ would not be a valid configuration.

**Definition 6.** A global transition $t_1=(b,c)_a$ is *maximal* if (i) $b$ and $c$ are maximal configurations, and (ii) no other global transition $t_2=(b,e)_a$ exists such that $m(t_1) \subset m(t_2)$.

Intuitively, $t_1$ pairs maximal configurations $b$ and c by firing a *maximal* set of nonconflicting TLChart transitions that are enabled in configuration $b$ using the inputs symbols in $a$. For example, in Figure A.1, ({*OPN, Wait-For-KeyPressed, State-3*}, {*CLS, Wait-For-KeyPressed, State-4*} $_{\{doorClosed, end\}}$ is a maximal transition but ({*OPN, Wait-For-KeyPressed, State-3*}, {*CLS, Wait-For-KeyPressed, State-3*} $_{\{doorClosed, end\}}$ is not maximal because it did not fire the enabled TLChart transition *State-3*$\rightarrow_{end}$ *State-4*.

ENFA transitions extend maximal transitions in that they might affect states in an indirect way, i.e., without those states being source or target states of any member transition. To this end, we need a few more definitions.

Recall that a statechart (are therefore a TLChart) default state is a designated initial state within a particular level of nesting. Hence, we define the following.

**Definition 7.** A state $d$ *is default under state s* if $d$ is a descendant of $s$ in the state tree, $d$ is a default state, and all ancestors of $d$ under $s$ are default states. A state $d$ *is a default state under a configuration C* if $d$ is a default state under the least common ancestor (lca) of the elements of $C$ in the state tree.

In Figure A.2 for example, state *Closed* is default under state *Door Thread*, but not default under the TLChart root. Finally, we define ENFA transitions.

**Definition 8.** Let $t=(b,c)_a$ be a maximal transition and let $b'$, $c'$ be the source and target sets of $t$, respectively. $t$ is a *legal ENFA transition* if for every state $s \in c$-$c'$ (i.e., $s$ is in $c$ but is not an explicit target state of some member transition), $s$ is a default state under $c$ or else $s \in b$-$b'$ (i.e., $s$ was not affected by the transition).

For example in Figure A.1, the maximal transition ({*Init*}, {*CLS, For-KeyPressed, State-3*} $_{\{begin\}}$ is an ENFA transition where the least common ancestor of configuration {*CLS, Wait-For-KeyPressed, State-3*} is state *State-1*, and *CLS, For-KeyPressed,* and *State-3* are all defaults under *State-1*.

Note that conflicting simultaneously enabled ENFA transitions induce nondeterminism. This is the case in the TLChart of Figure A.2 when *keyPressed, doorOpen,* and *alarm* are all true while in configuration {*Closed, Wait-For-KeyPressed*} enabling two ENFA transitions, one resulting in configuration {*Open, Wait-For-Key-Pressed*} while the other results in configuration {*Error*}.

## A.4. Semantics with Temporal Conditions

While standard semantics for LTL are defined using infinite models, in this paper our primary interest is with finite linear model semantics. Hence, for example, *Eventually* $\rho$ is satisfied if there exists state *s* in the finite linear model that satisfies $\rho$.

An LTL model relates to an automaton's input tape in the following straightforward way. An LTL model consists of a finite sequence of states with Boolean propositions and corresponding truth assignments assigned to each state. For example, consider a model with two states (i.e., two cycles), where {*begin, ¬end, KeyPressed, ¬alarm, doorOpen*} is the truth assignment for state 0 (interpreted as cycle 0), and {*¬begin, ¬end, KeyPressed, ¬alarm, doorClosed*} is the truth assignment for state 1. This model is, therefore, obviously exchangeable with an automaton input tape with the symbol *<begin, ¬end, KeyPressed, ¬alarm, doorOpen>* in position 0 and *<¬begin, ¬end, KeyPressed, ¬alarm, doorClosed>* in position 1. In other words, each Boolean proposition $p_i$ and its negation $\neg p_i$ form an alphabet $\Sigma_i$. The input alphabet $\Sigma$ for the ENFA is then the Cartesian product of all $\Sigma_i$ alphabets.

We now incorporate temporal conditions into ENFA behavior. First, note that every ENFA transition has a pair of propositional and temporal conditions, which are the respective conjunctions of all propositional and temporal conditions annotating its member TLChart transitions. Hence we represent an ENFA transition as an annotated binary relation $(b,c)_{a,\rho}$ where $a$ is a propositional condition and $\rho$ is a temporal condition. Temporal conditions affect ENFA behavior via the definition of a computation. Given an input tape, a conventional one-way nondeterministic Finite Automaton (NFA) computation is essentially a sequence of consecutive transitions and corresponding tape head moves to the right, as described in Chapter 1. ENFAs extend this well-known definition by requiring that for every transition $t_i$ in the computation the input tape is observed from position $i$ into the future and back to the past, but without moving the tape head. The transition $t_i$ is then enabled only if the temporal condition is satisfied by the tape, while considering position number $i$ as cycle 0.

**Definition 9.**  Let $\sigma=\sigma_1.\sigma_2...\sigma_n$ be an input tape and let $C=c_0.c_1...c_n$ be an ENFA computation. $C$ is a *computation on* $\sigma$ if $c_0$ is an initial configuration and $\forall i$, $0<i\le n$, the ENFA contains a transition $(c_{i-1},c_i)_{a,\rho}$ such that $a\subseteq\sigma_i$ and $\sigma,i\models\rho$ using standard LTL and MTL semantics (e.g., [CPM]).

For example, using the entry system TLChart of Figure A.2, consider the input tape (using straightforward abbreviations of the entry system conditions): $\sigma=\sigma_1.\sigma_2.\sigma_3.\sigma_4.\sigma_5.\sigma_6=$

$\{B,\neg E,KP,\neg A,DC\}.\{\neg B,\neg E,\neg KP,\neg A,DC\}.$

$\{\neg B,\neg E,KP,\neg A,DO\}.\{\neg B,\neg E,KP,A,DC\}.$

$\{\neg B,E,\neg KP,\neg A,DC\}.\{B,\neg E,\neg KP,\neg A,DO\}.$

The following $C_1$ computation is enabled by $\sigma$; each line is considered as a cycle, starting at cycle 0:

$\{Init\} \rightarrow_B$

$\{Wait\text{-}For\text{-}KeyPressed, Closed\} \rightarrow_{(none)}$

$\{Wait\text{-}For\text{-}KeyPressed, Closed\} \rightarrow_{KP, DO}$

$\{Wait\text{-}For\text{-}KeyPressed, Open\} \rightarrow_A$

$\{Error\}$ (a sink state)

Similarly, the following $C_2$ computation is also enabled by $\sigma$:

$\{Init\} \rightarrow_B$

$\{Wait\text{-}For\text{-}KeyPressed, Closed\} \rightarrow_{(none)}$

$\{Wait\text{-}For\text{-}KeyPressed, Closed\} \rightarrow_{KP, DO}$

$\{Wait\text{-}For\text{-}KeyPressed, Open\} \rightarrow_{KP,DC}$

$\{Wait\text{-}For\text{-}KeyPressed, Closed\} \rightarrow_\rho \{Done\}$

where $\rho$ is the temporal condition $E \ U \ B$. $\rho$ is enabled on cycle 4 because the input tape then points to $\sigma_5$ and the tape suffix is $\sigma_5.\sigma_6 =$

$\{\neg B, E, \neg KP, \neg A, DC\}.\{B, \neg E, \neg KP, \neg A, DO\}$ which satisfies $\rho$.

Like their logical counterpart, ENFAs represent assertions about a system. They do so using a formal language mechanism, namely by accepting or rejecting strings (tapes). A classical NFA accepts a string using an existential criterion, namely, if a computation ending in a final state exists. A dual universal automaton ($\forall$-FA) accepts a string if all computations end in a final state. Combining both acceptance criteria results in an alternating automaton. Alternatively, an existential NFA with negation can be used instead of a combination of both acceptance criteria. ENFA supports negation using (i) negation inside temporal conditions, (ii) a combination of *good* (accepting) and *error* (rejecting) states. For a given input string $s$, an ENFA has one or more possible computations on $s$, some of which end in a good state while others end in an error state. Conflicts are resolved using a priority scheme where the winning computation is the computation whose last visited state configuration contains a TLChart state *St* whose priority is higher than all other TLChart states in all competing configurations. If *St* is a good state then the TLChart accepts the input string, otherwise the TLChart rejects it. For example, in Figure A.2 consider two computations on the input string $\sigma$ described earlier, $C_1$ and $C_2$. $C_1$ ends in the configuration {*Error*} where the *error* state *Error* has priority 2. $C_2$ ends in the configuration {*Done*} where *good* state *Done* has priority 1. We use the engineering convention where lower integer values represent higher priorities; hence, $\sigma$ is accepted because *Done* has a higher priority than *Error*.

Whenever the priority scheme cannot resolve conflicts, we arbitrarily select the error computation as overriding. Likewise, whenever a single computation ends in a configuration that contains both good and error states, then we arbitrarily select the error state as overriding.

TLCharts support methods for real-time constraint specification. The first method uses Harel statechart *timeout* (*tm*) events, not unlike the mechanism used by timed automata. The second method uses MTL; consider a variant of Figure A.2 where the pair of transitions *Wait-For-KeyPressed*$\rightarrow_{tm(2min)}$*Alarm-Necessary*, and *Wait-For-KeyPressed*$\rightarrow_{keyPressed}$ *Wait-For-KeyPressed* are replaced with a single transition *Wait-For-KeyPressed*$\rightarrow_{\rho}$*Alarm-Necessary*, where $\rho = \neg\Diamond_{\leq 2min}$ *keyPressed*. Though similar, the two approaches differ with respect to the timing in which state *Alarm-Necessary* is reached. With the first representation, *Alarm-Necessary* is reached after two minutes, while under the second representation, the transition is traversed after one cycle. In this context, we suggest a special visual delay construct, represented with thick edges, which can only be used with the following unnested temporal conditions: $[]_{\leq d}\rho$ ($[]\rho$ with an MTL upper bound $d$), $\Diamond\rho$, and $\rho U \psi$. It means that the transition is traversed only when the temporal condition becomes true, i.e., when the MTL upper bound $d$ in for $[]_{\leq d}\rho$ is reached, or when $\psi$ is true in $\Diamond\psi$ or $\rho U \psi$. Hence, in Figure A.2, the transition *Done*→ $_{alarm}Error$ is enabled only after the final *keyPressed* input that satisfies the preceding transitions' temporal condition (*alarm U keyPressed*) is detected. A thick transition $A \rightarrow_{\rho U \psi} B$ is formally but a shorthand representation of a pair of *thin* transitions $A \rightarrow_{\rho U O \psi} A'$ and $A' \rightarrow_{\psi} B$, where A' is a new sibling of A in the state tree. A similar approach is used for the other types of thick transitions.

From a semantic perspective, real-time measurements, used by statechart timeout events and MTL constraints, are represented in our ENFA model using a standard monotonically increasing positive integer function that maps each tape cell with a real-time value.

Note that though visually similar to Harel Statecharts, TLCharts are actually used and applied more like a temporal logic specification in the following sense. TLCharts do not describe the token-by-token reaction of a reactive system to environment stimuli. Rather, TLCharts consider a complete input string $s$, which combines both environment inputs $s\_in$ and system outputs $s\_out$; a TLChart asserts about the legality of an $s\_out$ system response to the $s\_in$ stimuli.

## A.5. TLCharts with Overlapping States

The proposed automata theoretic statechart semantics described in Section A.3 caters to statecharts with overlapping states [Ka].

Consider the TLChart of Figure A.4, a variant of the TLChart of Figure A.3 but with overlapping states. In Figure A.4, state *State-OVLP* is an *and* state that shares its substates with the concurrent threads of state *State-2*. Figure A.4 induces the DAG state graph of Figure A.5. The intuitive meaning of the state overlap in Figure A.4 is that it is illegal for a key to be pressed while the valve is open.

From a semantics perspective, ENFA state configurations for TLCharts with overlapping states contain all state nesting information. Hence, the situation where *State-KP* and *Open* are simultaneously visited has two distinct possible representations as ENFA state configurations: {*State-1, State-2, State-KP, Open*}, and {*State-1, State-OVLP, State-KP, Open*}. Therefore, the following two computations are distinct, though when considering only leaf states, the cycle #1 configurations look alike:

$\{Init\} \rightarrow_B$

$\{State\text{-}1, State\text{-}2, Wait\text{-}For\text{-}KeyPressed, Closed\} \rightarrow_{KP,DO}$

$\{State\text{-}1, State\text{-}2, State\text{-}KP, Open\} \rightarrow_{DC}$

$\{State\text{-}1, \mathbf{State\text{-}2}, State\text{-}KP, Closed \}$

and

$\{Init\} \rightarrow_B$

$\{State\text{-}1, State\text{-}2, Wait\text{-}For\text{-}KeyPressed, Closed\} \rightarrow_{KP,DO}$

$\{State\text{-}1, \mathbf{State\text{-}OVLP}, State\text{-}KP, Open\} \rightarrow_{(any)}$

$\{Error\}$

Given that the second computation ends in *Error*, a state with higher priority than any of the states in {*State-1, State-2, State-KP, Closed*}, the TLChart rejects the input, effectively stating that *State-KP* and *Open* cannot be visited simultaneously.

Figure A.4 contains another instance of state overlapping, where two *or* states overlap, namely *NoAlarm* and *State-1*. Hence, when state *Wait-For-KeyPressed* is visited, it can be considered as residing under state *NoAlarm* or under state *State-1,* each case resulting in a different configuration. Therefore, the following two computations are possible:

$\{Init\} \rightarrow_B$

$\{State\text{-}1, State\text{-}2, Wait\text{-}For\text{-}KeyPressed, Closed\} \rightarrow_{alarm}$

$\{State\text{-}1, State\text{-}2, Wait\text{-}For\text{-}KeyPressed, Closed\}$

and

$\{Init\} \rightarrow_B$

$\{NoAlarm, Wait\text{-}For\text{-}KeyPressed\} \rightarrow_{alarm}$

$\{Error\}$

The formal semantics described earlier is extended to support TLCharts with overlapping states in the following way. Given a TLChart $T$ with overlapping states, we say that *T's state DAG contains the tree Tr*, if *Tr* is a spanning tree *in T and is a legal state-tree.* Using the existing semantics of Section A.3, every state-tree $ST_i$ contained in the state DAG induces an ENFA, which in turn induces a set $S_i$ of possible, competing computations; the final outcome of these competing computations is then determined using a priority scheme. The new extended semantics defines the *DAG computation set* $S_{DAG}$ as the union of all $S_i$ sets, and the final outcome is then determined using the priority scheme applied to all computations in $S_{DAG}$.

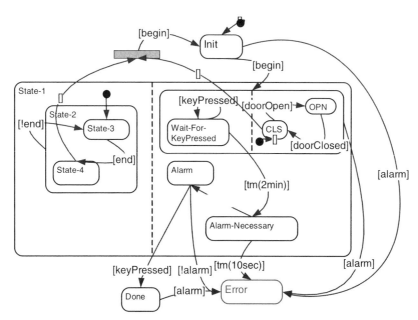

FIGURE A.1   Deterministic Harel statechart specification for requirement
R. A Harel statechart is by definition also a TLChart.

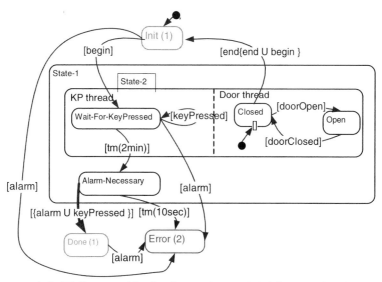

FIGURE A.2   TLChart specification for requirement R1 (all states other than Error
are good states; all states with no specified priority have default, i.e., lowest, priority).

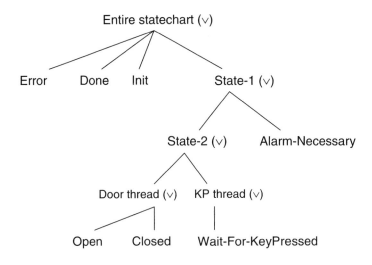

FIGURE A.3   State and/or tree for Figure A.2.

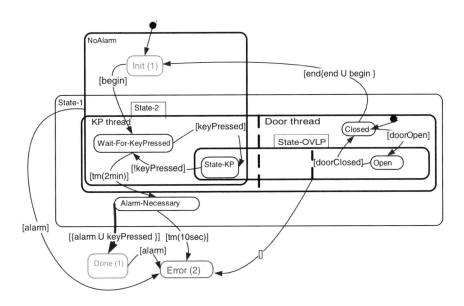

FIGURE A.4   A TLChart with overlapping states.

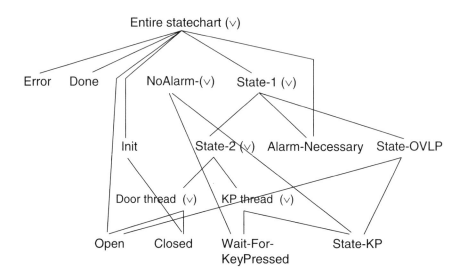

FIGURE A.5   State and/or DAG for the TLChart of Figure A.4.

# Notes

# Bibliographical Notes

Hopcroft, Motwani, and Ullman's book on automata and formal languages [HMU] is an excellent resource used by many schools across the country for teaching the theoretical background discussed briefly in Chapter 1. It includs formal definitions (albeit using traditional, academic, notation and vocabulary, which differ from the notation and vocabulary used in this book), as well as many interesting details not discussed in this book, such as a minimization algorithm for DFA, a formal subset construction algorithm, and lower bound proofs for NFA to DFA conversions.

Kohavi [Ko] describes a minimization technique for FSMs. Drusinsky provided a minimization technique for HFSM's [D1].

Sipser provided interesting succinctness lower bounds for NFA [Si]. Chandra, Kozen, and Stockmeyer introduced AFA in [CKS], and proved that AFA are as descriptive as DFA.

EREs are supported by programming languages such as Perl [Pe] and Java [O].

Rabin [Ra], Streett [Sa1], and Buchi [Bu] investigated infinite tape automata ($\omega$-DFA and $\omega$-NFA). See also [Wo]. Safra reports results for upper and lower bounds converting between various types of infinite-tape automata [Sa2]. Hayashi and Miyano invetstigated infinite tape AFA [HM].

Statecharts were introduced by Harel in [Ha1]. They were later introduced to the OMT methodology by Rumbough et al., [Ru] and then were made an integral part of the UML standard. Drusinsky introduced TLCharts in [D2]. A short version of this paper is available in the appendix. Harel et al. published a formal semantics for Harel statecharts in [Ha2]. Drusinsky and Harel introduced the first implementation technique for statecharts [DH1] referred to in this book as the horizontal method and also investigated the succinctness of statecharts [DH2].

According to Wikipedia [Wi], Temporal Logic was first studied in depth by Aristotle. In its contemporary form it was first introduced by Arthur Prior in a philosophical context [Pr] and then brought into the forefront of computer aided verification research by Amir Pnueli, the 1996 Turing award recipient. Together with Zohar Manna, he is the author of a three-volume textbook on Temporal Logic and its application to Reactive Systems of which the first two volumes are [MP1] and [MP2].

Koymans introduced Metric Temporal Logic [Koi]. Drusinsky introduced the concept of Runtime Execution Monitoring of LTL and MTL specifications [D3]. Havelund and Rosu developed a monitoring engine based on rewriting rules [HR].

Thomas proved that LTL's descriptive power is equivalent to that of star-free regular expressions [To]. Markey showed that temporal logic with past time is exponentially more succinct than future time LTL.

The genealogy of Tableau methods of converting LTL to $\omega$-NFA (Buchi automata) is described in [Tab].

Electronic Design Automation (EDA) Accellera organization (used for the verification of hardware designs) includes a language extension named PSL with support for ERE and LTL [Ac].

[Bra] describes various forms of branching time temporal logic. ITL was first described by Moszkowski [Mo]. **Melliar-Smith** introduced GIL in [zzz].

Propositional logic assertions are supported by the Java [JavaA] platform as well as some C and C++ versions. The use of propositional logic to capture preconditions, postconditions, and invariants in for transformational systems is described in [Ei] for Eiffel and in [JML] for JML.

Clarke, Grumberg, and Peled describe the state of the art model checking theory and practice in [CGP].

Holzmann describes the SPIN model checker in [Ho]. Wisser and Havelund describe the JPF execution-based model checker in [Wiz].

Drusinsky describes a specification based automatic test generator for LTL in [D3], and a two-level runtime recovery technique in [D4].

The Kansas State specification pattern library is described in [Ka1] and is available from [Ka2].

---

[Ac] www.accellera.org.

[Bra] Penczek W., "Branching Time and Partial Order in Temporal Logics." In *Time and Logic: A Computational Approach*, Univ. College of London, 1995, pp.179–228.

[Bu] J.R. BÄuchi. "On a decision method in restricted second order arithmetic." In *Proc. Internat. Congr. Logic, Method. and Philos. Sci.* 1960, pp. 1–12, Stanford, 1962. Stanford University Press.

[CKS] A. K. Chandra, D. Kozen, and L. J. Stockmeyer. "Alternation." In *Journal of the Association for Computing Machinery*, 28:114–133, 1981.

[CGP] *Model Checking*, Edmund M. Clarke, Orna Grumberg and Doron A. Peled, the MIT Press, January 2000, ISBN 0-262-03270-8.

[D1] D. Drusinsky, "Symbolic cover minimization of fully I/O specified finite state machines." In *Computer-Aided Design of Integrated Circuits and Systems, IEEE Transactions,* Vol. 9, Issue 7, Jul. 1990, pp. 779–781.

[D2] D. Drusinsky, "Semantics and Runtime Monitoring of TLCharts: Statechart Automata with Temporal Logic Conditioned Transitions." Electronic Notes in *Theoretical Computer Science* Copyright © 2006, Elsevier B.V. All rights reserved, Volume 113, pp. 1–233 (3 January 2005) Proceedings of the Fourth Workshop on Runtime Verification (RV 2004). 20040403-20040403. Edited by K. Havelund, Roşu.

[D3] D. Drusinsky. *The Temporal Rover and the ATG Rover*. Proc. Spin2000 Workshop. Springer Verlag Lecture Notes in Computer Science, 1885, pp. 323–329.

[D4] Drusinsky, Doron, "Specs Can Handle Exceptions." In *Embedded Developers Journal*, November 2001, p. 8.

[DH1] D. Drusinsky and David Harel, "Using Statecharts for Hardware Description and Synthesis." In *IEEE Transactions on Computer-Aided Design of Integrated Circuits and Systems* 8(7): 79–806, July 1989. Also appeared in Proc. ICCAD, 1988.

[DH2] D. Drusinsky and David Harel. "On the power of bounded concurrency I: Finite Automata." In *Journal of the ACM*, 41(3): 517–539, May 1994.

[Ei] *Object-Oriented Software Engineering with Eiffel* By Jean-Marc Jézéquel, from the PAMPA Project, Addison-Wesley, Eiffel in Practice Series, ISBN 0-201-63381-7, 1996.

[HMU] John E. Hopcroft, Rajeev Motwani, Jeffrey D. Ullman, *Introduction to Automata Theory, Languages, and Computation*, 2nd Edition. Addison Wesley, June 2001, ISBN: 0201441241.

[Ha1] D. Harel, "Statecharts: A Visual Approach to Complex Systems," *Science of Computer Programming*, Vol. 8, No. 3, pp. 231–274, 1987.

[H2] D. Harel and A. Naamad, *The STATEMATE semantics of statecharts, ACM Transactions on Software Engineering and Methodology (TOSEM),* archive Volume 5 , Issue 4 (October 1996) table of contents, pp. 293–333. Year of Publication: 1996.

[HR] Havelund, K. and Rosu, G. "Monitoring Java Programs with Java PathExplorer," in *Proc. of the First Workshop on Runtime Verification (RV'01)*. 23 July 2001, Paris, France, Electronic Notes in Theoretical Computer Science 55(2).

[HM] S. Miyano and T. Hayashi. "Alternating Finite automata on infinite-words." In *Theoretical Computer Science*, 32:321{330, 1984.

[Ho] G.J. Holzmann: *SPIN Model Checker, The Primer and Reference Manual* (see following) ISBN: 0-321-22862-6, Publisher: Addison Wesley Professional, 2004.

[JavaA] http://java.sun.com/developer/Books/javaprogramming/jdk14/javapch06.PDF.

[JML] http://www.cs.iastate.edu/~leavens/JML/index.shtml.

[Ka1] *Property Specification Patterns for Finite-state Verification*, Matthew B. Dwyer, George S. Avrunin and James C. Corbett in the 2nd Workshop on Formal Methods in Software Practice, March, 1998. An abstract of this paper is also available.

[Ka2] http://patterns.projects.cis.ksu.edu/documentation/patterns.shtml

[Ko] Z. Kohavi, *Switching and Finite Automata Theory*. New York, McGraw-Hill, 1978.

[Koi] R. Koymans. *Specifying Real-Time Properties with Metric Temporal Logic*. Real-time Systems, 2(4):255–299, 1990.

[Ma] Nicolas Markey. "Temporal logic with past is exponentially more succinct," Concurrency Column. *Bulletin of the EATCS* 79: pp. 122–128 (2003).

[Mo] Ben Moszkowski, *Executing Temporal Logic Programs*, Cambridge University Press, New York, NY, 1986.

[MP1] Z. Manna & A. Pnueli: *The Temporal Logic of Reactive and Concurrent Systems: Specification*, Springer-Verlag, 1991.

[MP2] Z. Manna & A. Pnueli: *Temporal Verification of Reactive Systems: Safety*, Springer-Verlag, 1995.

[M] P. M. Melliar-Smith. "A graphical representation of interval logic." In *Proc. Inter. Conf. Concurrency*, pp. 106–120, Hamburg, FRG, Oct. 1988. LNCS 335, Springer-Verlag.

[O] *An Introduction to Object-Oriented Programming with Java 1.5*. ISBN: 0073043915. Author: C. Thomas Wu (Otani). Publisher: McGraw-Hill Science/Engineering/Math Edition: 3. Date published 2004-05-28.

[Pe] Randal Schwartz and Tom Phoenix and Brian D. Foy, *Learning Perl* 4th edition. Release date 01 Jul 2005. Publisher O'Reilly & Associates. ISBN 0-596-10105-8. Randal Schwartz and Tom Phoenix and Brian D. Foy.

[Pr] http://www.kommunikation.aau.dk/prior/index2.htm

[Ra] M.O. Rabin. "Automata on Infinite Objects and Church's Problem." *Conference Series in Mathematics*, Vol. 13. AMS, 1969.

[Ru] *Object-Oriented Modeling and Design* (Hardcover) by James R Rumbaugh, Michael R. Blaha, William Lorensen, Frederick Eddy, William Premerlani. Publisher: Prentice Hall; 1st edition (October 1, 1990). ISBN 0136298419.

[Sa1] S. Safra. "On the complexity of ω-automata." In *29th annual Symposium on Foundations of Computer Science*, October 24–26, 1988, White Plains, New York, pp. 319–327. IEEE Computer Society Press, 1988.

[Sa2] S. Safra. "On the complexity of !-automata." In *Proc. 29th IEEE Symp. on Foundations of Computer Science*, pp. 319–327, White Plains, October 1988.

[Si]M. Sipser, "Lower bounds on the size of sweeping automata." In *Annual ACM Symposium on Theory of Computing*, 1979, pp. 360–364.

[T] "A combinatorial approach to the theory of !-automata." In *Information and Computation*, 48:261–283, 1981.

[Tab] http://spot.lip6.fr/wiki/LtlTranslationAlgorithms

[Wiz] "Program Model Checking as a New Trend," K. Havelund and W. Visser *International Journal on Software Tools for Technology Transfer (STTT)* Volume 4, Number 1, October 2002.

[Wi] http://en.wikipedia.org/wiki/Linear_temporal_logic

[Wo] Wolfgang Thomas, "Automata on infinite objects." In Van Leeuwen, Ed., *Handbook of Theoretical Computer Science*, pp. 133–164, Elsevier, 1990.

# About the Author

Doron Drusinsky received his B.Sc. from the Technion, Israel Institute of Technology in 1983, and received his PhD from Israel's Weizmann Institute in 1988. He worked for Sony's semiconductor division in Japan and in the U.S. from 1988–1993. He developed BetterState, the first PC-based statechart development tool. The tool was acquired by Integrated Systems (now part of Wind River Systems), in 1997. He then founded Time Rover Inc., and developed the TemporalRover, DBRover, ATGRover and StateRover tools. Doron joined the faculty of the Naval Postgraduate School, Monterey, CA, in 2002 and teaches computer science and software engineering.

# Index

## A

alternating finite automata 19, 21, 299
assertion statecharts 141–143, 145, 150, 163, 194–195, 200
assertions 104, 141
  applying assertions 229
  assertion actions 146
  assertion types 232–233
  bounded liveness assertions 155
  chaining assertions 159–160
  flag raisers 161–162, 196
  assertion libraries 240–241
  nesting assertions 158–159
  role of assertions during REM 244
  scoping assertions 157–158
  simulating and testing assertions 147
  simulating assertions 126, 189, 239, 261
  violations 201–202
  writing assertions 230–231
automatic test generation (ATG) 247

## B

battle management command and control (BMC2) 263
black box testing 99
black-box execution-based model checking 251–252
Boolean 14, 39, 49, 62–63, 91, 108–109, 122–123, 125, 132, 139, 141–142, 145, 147, 155–156, 196, 206, 212, 236, 243–244, 246, 256, 285

## C

code generation 44, 50, 57, 72–77, 79–81, 83, 87–88, 125, 139, 169, 171
  flat implementation 76
  hierarchical method 73
  horizontal technique 74
  robust code generator 50, 83
  vanilla code generator 60, 66, 80–83
  vertical method 76
compound computation-object 27
compound-object 26, 28, 31–32
computation (a Run) 10–11
computation-object 12–13, 16, 18–19, 21, 23, 26–28, 32, 37
present-state 12–13
computation tree 15–16, 18, 21–23, 26–27, 138
  computation tree trace 21
  leaves 16
computation tree logic (CTL) 138
computation paths 16
concurrence 44, 58, 60–61, 63–66, 163–164, 209–210, 281, 283
condition 1, 3, 23, 48–49, 58, 62–63, 79, 94, 105, 108–109, 112, 116, 125, 132–133, 139, 163, 165, 206, 215, 256, 281, 286–287, 289
correctness property 104
critical regions 87, 92–93

## D

Department of Defense (DoD) 261–262, 264–265
deterministic finite automata 8–10, 163
deterministic FA (DFA) 9–19, 22, 24–32, 34–38, 40–41, 121–123, 162–163, 210, 295
  fully specified 10–11, 15
domain of discourse (alphabet) 1–2, 25, 47
  letters 2

## E

ε-transitions 22–25
Eclipse 45, 264
enumerated flowchart visual switch polygons 87, 91
equivalent nondeterministic automaton (ENFA) 282
event 1–4, 14–15, 23–24, 47, 49–52, 56–58, 60, 62–63, 65, 68, 71, 73, 76–84, 86–88, 90–94, 96–97, 105, 127, 131–133, 146–147, 150–151, 155–156, 158, 161–167, 170, 173–174, 179–183, 185–186, 188–191, 195–197, 200–201, 204, 206, 213–215, 223, 230–231, 241, 243, 254–257, 259, 273, 275, 307–308
event-driven 15, 51, 83–86, 150, 153, 155, 203, 206–207
event-driven statecharts 51, 83–84, 206–207

## ELSEVIER SCIENCE CD-ROM LICENSE AGREEMENT

PLEASE READ THE FOLLOWING AGREEMENT CAREFULLY BEFORE USING THIS CD-ROM PRODUCT. THIS CD-ROM PRODUCT IS LICENSED UNDER THE TERMS CONTAINED IN THIS CD-ROM LICENSE AGREEMENT ("Agreement"). BY USING THIS CD-ROM PRODUCT, YOU, AN INDIVIDUAL OR ENTITY INCLUDING EMPLOYEES, AGENTS AND REPRESENTATIVES ("You" or "Your"), ACKNOWLEDGE THAT YOU HAVE READ THIS AGREEMENT, THAT YOU UNDERSTAND IT, AND THAT YOU AGREE TO BE BOUND BY THE TERMS AND CONDITIONS OF THIS AGREE-MENT. ELSEVIER SCIENCE INC. ("Elsevier Science") EXPRESSLY DOES NOT AGREE TO LICENSE THIS CD-ROM PRODUCT TO YOU UNLESS YOU ASSENT TO THIS AGREEMENT. IF YOU DO NOT AGREE WITH ANY OF THE FOLLOWING TERMS, YOU MAY, WITHIN THIRTY (30) DAYS AFTER YOUR RECEIPT OF THIS CD-ROM PRODUCT RETURN THE UNUSED CD-ROM PRODUCT AND ALL ACCOMPANYING DOCUMENTATION TO ELSEVIER SCIENCE FOR A FULL REFUND.

### DEFINITIONS

As used in this Agreement, these terms shall have the following meanings:

"Proprietary Material" means the valuable and proprietary information content of this CD-ROM Product including all indexes and graphic materials and software used to access, index, search and retrieve the information content from this CD-ROM Product developed or licensed by Elsevier Science and/or its affiliates, suppliers and licensors.

"CD-ROM Product" means the copy of the Proprietary Material and any other material delivered on CD-ROM and any other human-readable or machine-readable materials enclosed with this Agreement, including without limitation documentation relating to the same.

### OWNERSHIP

This CD-ROM Product has been supplied by and is proprietary to Elsevier Science and/or its affiliates, suppliers and licensors. The copyright in the CD-ROM Product belongs to Elsevier Science and/or its affiliates, suppliers and licensors and is protected by the national and state copyright, trademark, trade secret and other intellectual property laws of the United States and international treaty provisions, including without limitation the Universal Copyright Convention and the Berne Copyright Convention. You have no ownership rights in this CD-ROM Product. Except as expressly set forth herein, no part of this CD-ROM Product, including without limitation the Proprietary Material, may be modified, copied or distributed in hardcopy or machine-readable form without prior written consent from Elsevier Science. All rights not expressly granted to You herein are expressly reserved. Any other use of this CD-ROM Product by any person or entity is strictly prohibited and a violation of this Agreement.

### SCOPE OF RIGHTS LICENSED (PERMITTED USES)

Elsevier Science is granting to You a limited, non-exclusive, non-transferable license to use this CD-ROM Product in accordance with the terms of this Agreement. You may use or provide access to this CD-ROM Product on a single computer or terminal physically located at Your premises and in a secure network or move this CD-ROM Product to and use it on another single computer or terminal at the same location for personal use only, but under no circumstances may You use or provide access to any part or parts of this CD-ROM Product on more than one computer or terminal simultaneously.

You shall not (a) copy, download, or otherwise reproduce the CD-ROM Product in any medium, including, without limitation, online transmissions, local area networks, wide area networks, intranets, extranets and the Internet, or in any way, in whole or in part, except that You may print or download limited portions of the Proprietary Material that are the results of discrete searches; (b) alter, modify, or adapt the CD-ROM Product, including but not limited to decompiling, disassembling, reverse engineering, or creating derivative works, without the prior written approval of Elsevier Science; (c) sell, license or otherwise distribute to third parties the CD-ROM Product or any part or parts thereof; or (d) alter, remove, obscure or obstruct the display of any copyright, trademark or other proprietary notice on or in the CD-ROM Product or on any printout or download of portions of the Proprietary Materials.

### RESTRICTIONS ON TRANSFER

This License is personal to You, and neither Your rights hereunder nor the tangible embodiments of this CD-ROM Product, including without limitation the Proprietary Material, may be sold, assigned, transferred or sub-licensed to any other person, including without limitation by operation of law, without the prior written consent of Elsevier Science. Any purported sale, assignment, transfer or sublicense without the prior written consent of Elsevier Science will be void and will automatically terminate the License granted hereunder.

## TERM

This Agreement will remain in effect until terminated pursuant to the terms of this Agreement. You may terminate this Agreement at any time by removing from Your system and destroying the CD-ROM Product. Unauthorized copying of the CD-ROM Product, including without limitation, the Proprietary Material and documentation, or otherwise failing to comply with the terms and conditions of this Agreement shall result in automatic termination of this license and will make available to Elsevier Science legal remedies. Upon termination of this Agreement, the license granted herein will terminate and You must immediately destroy the CD-ROM Product and accompanying documentation. All provisions relating to proprietary rights shall survive termination of this Agreement.

## LIMITED WARRANTY AND LIMITATION OF LIABILITY

NEITHER ELSEVIER SCIENCE NOR ITS LICENSORS REPRESENT OR WARRANT THAT THE INFORMATION CONTAINED IN THE PROPRIETARY MATERIALS IS COMPLETE OR FREE FROM ERROR, AND NEITHER ASSUMES, AND BOTH EXPRESSLY DISCLAIM, ANY LIABILITY TO ANY PERSON FOR ANY LOSS OR DAMAGE CAUSED BY ERRORS OR OMISSIONS IN THE PROPRIETARY MATERIAL, WHETHER SUCH ERRORS OR OMISSIONS RESULT FROM NEGLIGENCE, ACCIDENT, OR ANY OTHER CAUSE. IN ADDITION, NEITHER ELSEVIER SCIENCE NOR ITS LICENSORS MAKE ANY REPRESENTATIONS OR WARRANTIES, EITHER EXPRESS OR IMPLIED, REGARDING THE PERFORMANCE OF YOUR NETWORK OR COMPUTER SYSTEM WHEN USED IN CONJUNCTION WITH THE CD-ROM PRODUCT.

If this CD-ROM Product is defective, Elsevier Science will replace it at no charge if the defective CD-ROM Product is returned to Elsevier Science within sixty (60) days (or the greatest period allowable by applicable law) from the date of shipment.

Elsevier Science warrants that the software embodied in this CD-ROM Product will perform in substantial compliance with the documentation supplied in this CD-ROM Product. If You report significant defect in performance in writing to Elsevier Science, and Elsevier Science is not able to correct same within sixty (60) days after its receipt of Your notification, You may return this CD-ROM Product, including all copies and documentation, to Elsevier Science and Elsevier Science will refund Your money.

YOU UNDERSTAND THAT, EXCEPT FOR THE 60-DAY LIMITED WARRANTY RECITED ABOVE, ELSEVIER SCIENCE, ITS AFFILIATES, LICENSORS, SUPPLIERS AND AGENTS, MAKE NO WARRANTIES, EXPRESSED OR IMPLIED, WITH RESPECT TO THE CD-ROM PRODUCT, INCLUDING, WITHOUT LIMITATION THE PROPRIETARY MATERIAL, AN SPECIFICALLY DISCLAIM ANY WARRANTY OF MERCHANTABILITY OR FITNESS FOR A PARTICULAR PURPOSE.

If the information provided on this CD-ROM contains medical or health sciences information, it is intended for professional use within the medical field. Information about medical treatment or drug dosages is intended strictly for professional use, and because of rapid advances in the medical sciences, independent verification of diagnosis and drug dosages should be made.

IN NO EVENT WILL ELSEVIER SCIENCE, ITS AFFILIATES, LICENSORS, SUPPLIERS OR AGENTS, BE LIABLE TO YOU FOR ANY DAMAGES, INCLUDING, WITHOUT LIMITATION, ANY LOST PROFITS, LOST SAVINGS OR OTHER INCIDENTAL OR CONSEQUENTIAL DAMAGES, ARISING OUT OF YOUR USE OR INABILITY TO USE THE CD-ROM PRODUCT REGARDLESS OF WHETHER SUCH DAMAGES ARE FORESEEABLE OR WHETHER SUCH DAMAGES ARE DEEMED TO RESULT FROM THE FAILURE OR INADEQUACY OF ANY EXCLUSIVE OR OTHER REMEDY.

## U.S. GOVERNMENT RESTRICTED RIGHTS

The CD-ROM Product and documentation are provided with restricted rights. Use, duplication or disclosure by the U.S. Government is subject to restrictions as set forth in subparagraphs (a) through (d) of the Commercial Computer Restricted Rights clause at FAR 52.22719 or in subparagraph (c)(1)(ii) of the Rights in Technical Data and Computer Software clause at DFARS 252.2277013, or at 252.2117015, as applicable. Contractor/Manufacturer is Elsevier Science Inc., 655 Avenue of the Americas, New York, NY 10010-5107 USA.

## GOVERNING LAW

This Agreement shall be governed by the laws of the State of New York, USA. In any dispute arising out of this Agreement, you and Elsevier Science each consent to the exclusive personal jurisdiction and venue in the state and federal courts within New York County, New York, USA.